Henry George Vennor

Vennor's winter almanac and weather record

1877-78

Henry George Vennor

Vennor's winter almanac and weather record
1877-78

ISBN/EAN: 9783337257828

Printed in Europe, USA, Canada, Australia, Japan

Cover: Foto ©berggeist007 / pixelio.de

More available books at **www.hansebooks.com**

VENNOR'S ALMANAC,

AND

WEATHER RECORD

FOR

1877-8.

MONTREAL:

Whilst no man had the means of knowing anything about the weather, beyond his sight, or the "feeling of his own instruments," it was scarcely possible to foretell changes of importance at a distance, as well as on the spot; but now the case is exceedingly different. A daily glance at the published "Weather Reports," a recollection of their principal features during the few previous days, a look at the "[illegible]" at home, and an eye turned [illegible]

Printed and Electrotyped at the WITNESS Establishment, 33 to 37 St. Bonaventure Street, Montreal.

INTRODUCTION.

"HOW ARE YOU? FINE DAY!" is a fair sample of the customary greeting of the members of the Anglo Saxon race the world over, whether it rains, blows, hails, snows, or the sun shines aloft. An Egyptian would greet a friend under similar circumstances with "How goes the perspiration?" a Greek of the present time, with, "What do you do?" a Dutchman, "How do you fare?" a Chinaman, "Have you eaten your rice?" or "Is your stomach in good order?" a Russian, "How do you live on?" or the very familiar, "Devil take you;" an Arab, "God grant thee His favor;" a Turk, "Be under the care of God;" a Persian, "Is thy exalted condition good?" a Japanese, "Do not hurt me;" and a Burmese, kissing the friend, "Give me a smell." It appears from these illustrations that in the matter of salutation the Anglo-Saxon has almost a monopoly of the weather, and it is not wonderful that he has a special interest in knowing what weather is to come, and has ever endeavored, and will continue to endeavor, to peer into the approaching seasons. Mr. Vennor contends that this desire is not without reason, and that nature has given her students very good ground to work upon in endeavoring to foretell weather. He argues that as a shepherd knows from experience what kind of day the morrow will be, by indications which never lie, any one, if he had the same experience in years, could as surely foretell the character of the coming year by the one preceding it. The great difficulty met immediately on the threshold of this theory is, that while there are three hundred and sixty-five or three hundred and sixty-six days, as the case may be, in a year, in this degenerate age, there are but some three score years and ten in a lifetime, and therefore the experience necessary to foretell the seasons is very difficult to obtain. This may be overcome by keeping a record of each day's character. That such a record will be soon obtained there can be no doubt, and the result of constant thought and enquiry can have but one conclusion. Mr. Vennor does not put forward his predictions as prophecies, but simply as opinions based on certain facts, and with the gallant Admiral Fitzroy, the founder of the present extensive and valuable meteorological system, may say, "Certain it is, that although our conclusions may be incorrect, our judgment erroneous, the *laws of nature* and the signs afforded to man are invariably true. Accurate interpretation is the deficiency." This accurate interpretation Mr. Vennor holds may be attained to, and study with that result in view will be rewarded by the most satisfactory results.

But this year VENNOR'S ALMANAC is not confined to the foretelling of seasons. It will have a new interest from the large amount of information and weather lore it contains; Virgil, Shakespeare, Longfellow, Howard, Admiral Fitzroy, Doctor Loomis, Steinmetz, Howe, Butler and other observers have been made to contribute to it, and the information obtained as the result of their observations will prove of value to those wise enough to take advantage of the experience of others. A few weather proverbs and superstitious rhymes have also been given a place in this little volume, but it is to be hoped that they will not unsettle the minds of any. If such be likely in the case of any reader, he is referred to the words of one of the most painstaking collectors of these sayings, the poet Gay, who, as an antidote to his work, says:

> **"Let no such vulgar tales debase thy mind;**
> **Nor Paul, nor Swithin rule the clouds or wind."**

But still some of these fables are based on shrewd observations, and sometimes deserve more attention than they obtain.

The pictures and descriptions of the different clouds given will prove an interesting study, and supplemented by close observation will enable the student to intelligently "discern the face of the sky."

It is a pleasing fact to note that during the year, all over the country, there have been many who have utilized the memorandum pages of the Almanac for 1877, noting therein interesting and curious incidents, "weatherwise and otherwise." In this regard Andrew Steinmetz, in his "Manual of Weather Casts and Prognostications on Land and Sea," says: "Every pocket book should have blank pages headed 'The Weather,' for each month of the year. It is obvious that by a little study and attention any one may soon become weatherwise, at least sufficiently so for ordinary purposes."

It is hoped that by the yearly publication of this manual, a new interest will be given to one of the most interesting and generally useful of studies, and that Mr. Vennor's facts and theories, whether right or wrong, will have the effect of causing a further investigation into Nature's secrets which will result in the increased usefulness and intelligence of the learner, who will be impressed the more he studies with the wisdom of Him "who doeth all things well."

ECLIPSES IN 1878.

In the year 1878 there will be four eclipses—two of the sun and two of the moon.

The first will be an annular eclipse of the sun, on February 2nd ; not visible from Canada.

The second will be a partial eclipse of the moon, on February 17th ; partly visible in Canada. It will begin at Halifax, N. S., 4h. 27m. in the morning, and at Montreal 3h. 47m. It will end at Halifax 9h. 25m., and at Montreal 8h. 46m.

The third will be a total eclipse of the sun, on the 29th July, visible in North America as a partial eclipse. It begins on the Earth 3h. 4m. evening, mean time at Halifax, in longitude 144° 50′ east of Greenwich, and latitude 41° 21′ north. Central Eclipse begins at 4h. 10m. evening, in longitude 117° 42′ east, and latitude 54° 14′ north. Ends on the Earth 8h. 1m. evening, in longitude 69° 45′ west, and latitude 3° 37′ north. This Eclipse begins at Halifax about 5h. 30m. evening, and at Montreal about 5h. 10m. The greatest observations will occur a few minutes before the setting of the sun.

The fourth will be a partial eclipse of the moon, on August 12th, and will be visible in part. In Halifax it begins in the evening at 5h. 8m., and in Montreal at 4h. 37m. The moon rises about 7 o'clock, and the last contact with the shadow will be about 9.

A transit of Mercury will occur on the 8th of May. It will be visible between 11 in the morning and 9 in the evening.

CHRONOLOGICAL CYCLES.—Dominical Letter, F ; Golden number, 16 ; Jewish Lunar Cycle, 14 ; Epact or Moon's Age, 16 ; Solar Cycle, 11 ; Julian Period, 6591 ; Jewish Year, commencing 30th September, 5638 ; Roman Indictions, 6 ; Mohammedan Year, 1395.

MOVABLE FESTIVALS.—Septuagesima Sunday, February 17th ; Sexagesima Sunday, February 23rd ; Quinquagesima Sunday, March 3rd ; Ash Wednesday, March 6th ; First Sunday in Lent, March 10th ; Mid-Lent Sunday, March 31st ; Palm Sunday, April 14th ; Good Friday, April 10th ; Easter Sunday, April 21st ; Low Sunday, April 28th ; Rogation Sunday, May 26th ; Ascension Day, May 30th ; Whitsunday, June 9th ; Trinity Sunday, June 16th ; Corpus Christi, June 20th ; Advent Sunday, December 1st.

HOLIDAYS OBSERVED IN PUBLIC OFFICES.—Circumcision, Jan. 1st ; Epiphany, January 6th ; Annunciation Virgin Mary, March 25th ; Good Friday, April 19th ; Ascension Day, May 30th ; Queen's Birthday, May 24th ; Corpus Christi, June 20th ; St. Peter and St. Paul, June 29th ; All Saints Day, November 1st ; Conception of the Blessed Virgin Mary, December 8th ; Christmas Day, December 25th.

BANK HOLIDAYS IN ONTARIO.—Sundays, Christmas Day, New Year's Day, Ash Wednesday, Good Friday, Easter Monday, The Queen's Birthday, and each day appointed by Royal Proclamation as a general Fast or Thanksgiving day.

CONTENTS.

INDEX TO ADVERTISEMENTS.

CHARLES PUNCHARD. Toronto and Montreal. CANVASSER FOR ADVERTISING.

MOON'S PHASES.

New Moon....	3rd	9.10 mo.	Full Moon.....	18th	7.17 ev.
First Quarter..	11th	1.53 ev.	Third Quarter..	25th	10.56 mo.

WEATHER PROVERBS AND WEATHER WISDOM.

1 Tues
2 Wed
3 Thu
4 Fri
5 Sat

If New Year's eve night-wind blow *south*,
It betokeneth warmth and growth ;
If *west*, much muck and fish in the sea ;
If *north*, much cold and storms there will be ;
If *east*, the trees will bear much fruit,
If *north-east*, flee it man and brute.

6 **Sun** **EPIPHANY—Twelfth Day.**

7 Mon
8 Tues
9 Wed
10 Thu
11 Fri
12 Sat

Catgut shrinks in wet weather, and thus the strings of violins and guitars shrink and snap before rain. The old-fashioned weathervane consisted of a man and woman so fixed before a house that by the contraction of the catgut on the approach of wet weather, the woman entered it, while

13 **Sun** **1st Sunday after Epiphany.**

14 Mon
15 Tues
16 Wed
17 Thu
18 Fri
19 Sat

a man, wrapped in a great coat, came out. When the weather grew fine, the woman came out and the man hid himself again. If a line of good well-dried whip cord and a plummet be hung against the wall and a mark made just where the plummet reaches, the plummet in very moderate weather

20 **Sun** **2nd Sunday after Epiphany.**

21 Mon
22 Tues
23 Wed
24 Thu
25 Fri
26 Sat

will be found to rise above the mark before rain, and sink below it as the weather becomes fair.

If St. Paul's day (25th January) be fair and clear,
It does betide a happy year ;
But if it chance to snow or rain,

27 **Sun** **3rd Sunday after Epiphany.**

28 Mon
29 Tues
30 Wed
31 Thu

Then will be dear all kinds of grain.
If clouds or mists do dark the skie,
Great store of birds or beasts shall die.

— *Willsford's Nature's Scenes*

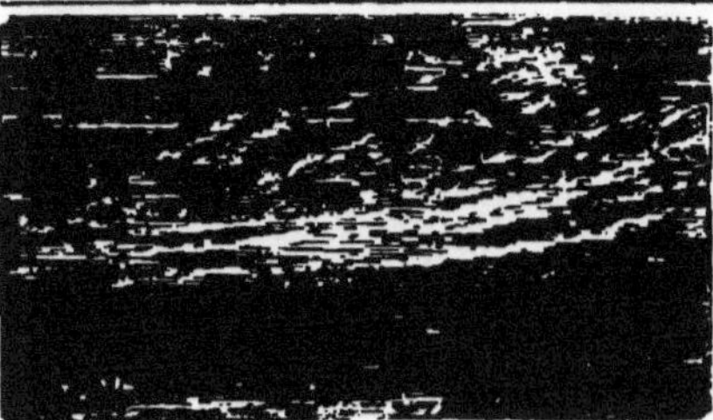

CIRRUS.

Cirrus Clouds consist of long, slender filaments, either parallel or divergent, which look like locks of hair ("mares tails"). They are generally very high up, sometimes covering the face of the sky with thin transparent gauze. If their under surface be horizontal, and their fibres point upwards, they indicate rain ; if downwards, fine weather, wind or drought. If cirrus clouds form during fine weather, with a falling barometer, it is almost sure to rain

HENDERSON'S LENDING LIBRARY,

191 St. Peter Street.

Henderson's Agency for Butterick's Cut Paper Patterns
191 St Peter Street.

Henderson's Magazine and News Depot,
191 St. Peter Street.

Henderson, the Bookseller and Stationer,
191 St. Peter Street.

For Scrap Pictures and Photographs, go to
HENDERSON'S.

For the largest and most select stock of Scripture, Birthday and Christmas Cards (especially the latter) held by any one house in the Dominion, go to HENDERSON'S.

For Children's Picture Books, for Children's Annuals, for Perforated Mottoes, for Illuminated Texts, for Book Marks, for the thousand and one items of a well regulated store, for a large assortment, goods fresh, and prices moderate, go to

J. HENDERSON'S

191 St. Peter St., (Branch 67 St. Lawrence St.,) Montreal.

EAGLE FOUNDRY.

14 to 34 King & Queen Streets, Montreal.

GEORGE BRUSH,

Maker of Steam Engines, Steam Boilers, Mill Machinery, Castings and Forgings of all kinds.

Agent for Heald & Sisco's Celebrated Pumps, Waters' Perfect Engine Governor, U. S. Hoisting and Conveying Co.'s Machine.

Specialties:
- **IMPROVED HAND & POWER ELEVATORS,**
- **PATENT BOILER TUBE BEADERS,**
- **WARRICK'S UNIVERSAL ENGINES.**

WHAT THEY SAY OF

DR. THOMAS' ECLECTRIC OIL.

MESSRS. NORTHROP & LYMAN are the sole agents for DR. THOMAS' ECLECTRIC OIL., which is now being sold in immense quantities throughout the Dominion. It is welcomed by the suffering invalid everywhere with emotions of delight, because it banishes pain and gives instant relief. This valuable specific for almost " every ill that flesh is heir too " is valued by the sufferer as more precious than gold. It is the elixir of life to many a wasted frame. If you have not purchased a bottle, do so at once, and keep it ready for an emergency. Its cheapness, 25 cents per bottle, places it within the reach of all. To the farmer it is indispensable, and it should be in every house.—*London Herald*

POOLS ISLAND, Nfld., Sept. 26th, 1876.

DEAR SIR,—I have been watching the progress of your ECLECTRIC OIL since its introduction to this place, and with much pleasure state that my anticipations of its success have been fully realized, it having cured me of bronchitis and soreness of nose ; while not a few of my "rheumatic neighbours" (one old lady in particular) pronounces it to be the best article of its kind that has ever been brought before the public. Your medicine does not require any longer a sponsor, but if you wish me to act as such, I shall be only too happy to have my name connected with your prosperous child. I am yours, &c.,

JAMES CULLEN.

W. W. McLellan, Lyn, P. O., N. S., writes :—"I was afflicted with rheumatism and had given up all hopes of a cure. By chance I saw Dr. Thomas' Eclectric Oil recommended. I immediately sent (fifty miles) and purchased four bottles, and with only two applications I was able to get around, and although I have not used one bottle I am nearly well. The other three bottles I gave around to my neighbours, and I have had so many calls for more that I felt bound to relieve the afflicted by writing to you for two dozen bottles."

P. M. Markell, West Jeddore, N. S., writes :—"I wish to inform you of the wonderful qualities of your Eclectric Oil. I had a horse so lame that he could scarcely walk ; the trouble was in the knee, and two or three applications completely cured him."

Five to thirty drops of THOMAS' ECLECTRIC OIL will cure common Sore Throat. It never fails in Croup. It will cure a cold or cough in twenty-four to forty-eight hours. One bottle has cured Bronchitis of eight years' standing ; recent cases are cured in three to six days. It has restored the voice where the person had not spoken above a whisper in five years. As an outward application in all cases of pain or lameness, nothing like it has ever been known. One bottle will cure any case of Lame Back or Crick in the Back. For diseases of the Spine and Contraction of the Muscles it is unequalled. In rheumatic or any other pain the first application does you good. It stops Ear Ache and the pain of a Burn in three minutes, and is altogether the best and cheapest medicine ever offered to the people.

Sold by all Medicine Dealers. Price 25 cents.

NORTHROP & LYMAN, Toronto, O., Proprietors for the Dominion.

TO LADIES.

A. ACKROYD,

MANUFACTURER OF

Corsets. Bustles and Human Hair Switches, Curls, Chignons, &c.,

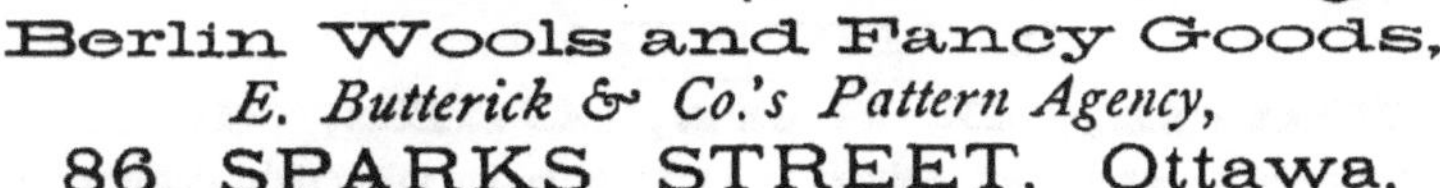

Berlin Wools and Fancy Goods,

E. Butterick & Co.'s Pattern Agency,

86 SPARKS STREET, Ottawa.

S. & H. BORBRIDGE,

WHOLESALE AND RETAIL MANUFACTURERS AND DEALERS IN

Harness, Saddles, Trunks, Valises, Carpet and Palissier Bags, Satchels, Horse Clothing, &c.,

Corner of Rideau and Mosgrove Streets, OTTAWA.

CANCER CURE.

OTTAWA, ONT.

TREATMENT WITHOUT THE KNIFE.

O. C. WOOD, Esq.. M.D.

DEAR DOCTOR,—You enquired after my health and views relative to your treatment of the cancer or cancerous affection in my lip—now just eleven years ago—in reference to which I have to express my gratitude to an over-ruling Providence that I was led to an acquaintance with you, and became a subject of treatment by you. My lip had been sore at least seven (7) years, exceedingly painful, and for two or three years before you took it in hand, almost unendurable. All sorts of experiments had been submitted to by me, embracing caustics, excoriation—everything indeed but the surgical knife · and in vain; for it always returned, and worse than before. Your treatment effected a speedy, complete and permanent cure. The cancerous humor seems thoroughly expurgated from my blood. I have now for a long time used nothing antagonistic as at first, nor any stimulant or tonic to keep up my system; and yet my health is perfect, and, at the age of 66, I am laboring with a vigor equal, if not superior, to any other part of my laborious life.

You are at liberty to make any use of this you may judge proper.

Yours gratefully,

JOHN CARROLL,

DON MOUNT, Ontario. *Wesleyan Methodist Minister.*

FEBRUARY.

MOON'S PHASES.

New Moon	2nd	3.24 mo.	Full Moon....	17th	6.23 mo.
First Quarter...	10th	8.23 mo.	Third Quarter.	23rd	10.19 ev.

1	FRI
2	SAT
3	**Sun**
4	MON
5	TUES
6	WED
7	THU
8	FRI
9	SAT
10	**Sun**
11	MON
12	TUES
13	WED
14	THU
15	FRI
16	SAT
17	**Sun**
18	MON
19	TUES
20	WED
21	THU
22	FRI
23	SAT
24	**Sun**
25	MON
26	TUES
27	WED
28	THU

WEATHER PROVERBS AND WEATHER WISDOM.

If February gives much snow,
A fine summer it doth foreshow.—*French Proverb.*

4th Sunday after Epiphany.

Above the rest, the sun, who never lies,
Foretells the change of weather in the skies ;
For if he rise unwilling to his race,
Clouds on his brow and spots upon his face,
Or if through mists he shoot his sullen beams,

5th Sunday after Epiphany.

Frugal of light, in loose and straggling streams,
Suspect a drizzling day, and southern rain,
Fatal to fruits, and flocks, and promised grain.—*Virgil.*

A dusty march, a snowy February, a moist April, and a dry May, presage a good year.—*French Proverb.*

Septuagesima Sunday.

Fine days in February are all borrowed from April.
—*Canadian Proverb*

February's nights, from the 22nd to the 28th are called in Sweden " steel nights" from their cutting severity.

Sexagesima Sunday.

St. Matthew breaks the ice ; if he finds none he will make it.

Before rain, flies cling to the ceiling or disappear ; spiders are restless, and frequently drop from the wall ; frogs croak importunately ; worms creep out of the ground ; bees cease to leave their hives, either remaining in them all day, or else flying only to a short distance.

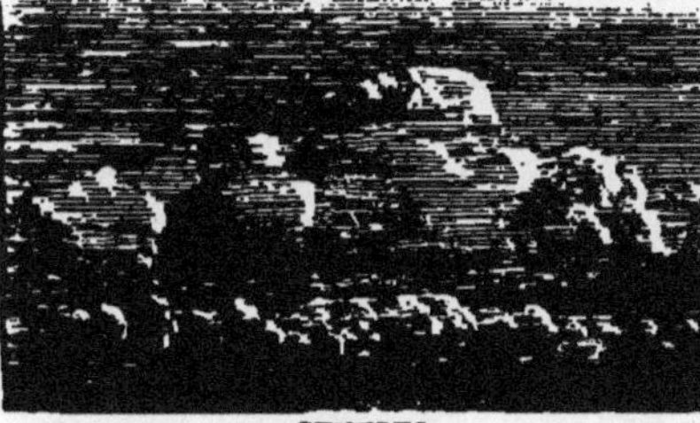

CUMULUS.

The *cumulus* cloud usually consists of a hemispherical or convex mass, rising from a horizontal base. It is much denser than the cirrus, and forms in the lower regions of the atmosphere. If it is fleecy and sails against the wind, it indicates rain ; but when the outline is very hard, and it comes up *with* the wind, it foretells fine weather. If cumulus clouds get smaller towards evening, expect fair weather ; if they increase at sunset, expect a thunderstorm at night.

MARCH.

MOON'S PHASES.

New Moon...	3rd	10.24 ev.	Full Moon.....	18th	4.13 ev.
First Quarter.	11th	11. 7 ev.	Third Quarter..	25th	11.56 ev.

		WEATHER PROVERBS AND WEATHER WISDOM.
1	FRI	March comes in with adders' heads, and goes out with
2	SAT	peacocks' tails.—*Scotch Proverb.*
3	**Sun**	**Quinquagesima—Shrove Sunday.**
4	MON	
5	TUES	When March is like April, April will be like March.
6	WED	A March without water dowers the king's daughter.
7	THU	—*French Proverbs.*
8	FRI	When March thunders, tools and arms get rusty.
9	SAT	—*Portuguese Proverb.*
10	**Sun**	**1st Sunday in Lent.**
11	MON	A peck of March dust, and a shower in May,
12	TUES	Make the corn green and the fields gay.—*English*
13	WED	*Proverb.*
14	THU	Sultriness, or the oppressive feeling we sometimes
15	FRI	experience, shows that the air is very damp ; it is already
16	SAT	
17	**Sun**	**2nd Sunday in Lent.**
18	MON	*saturated* with vapor, and therefore the evaporation from
19	TUES	our bodies is checked ; hence the oppressive feeling.
20	WED	The three first days of March, (old style) are called the
21	THU	borrowing days, for as they are remarked to be unusually
22	FRI	stormy, it is feigned that March had borrowed them from
23	SAT	April to extend the sphere of his rougher sway.—*Note in*
24	**Sun**	**3rd Sunday in Lent.** [*"Heart of Mid Lothian."*
25	MON	
26	TUES	March borrowit from April
27	WED	Three days, and they were ill ;
28	THU	The first was frost, the second was snow,
29	FRI	The third was cauld as ever't could blaw.—*Scotch*
30	SAT	*Proverb.*
31	**Sun**	**4th Sunday in Lent.**

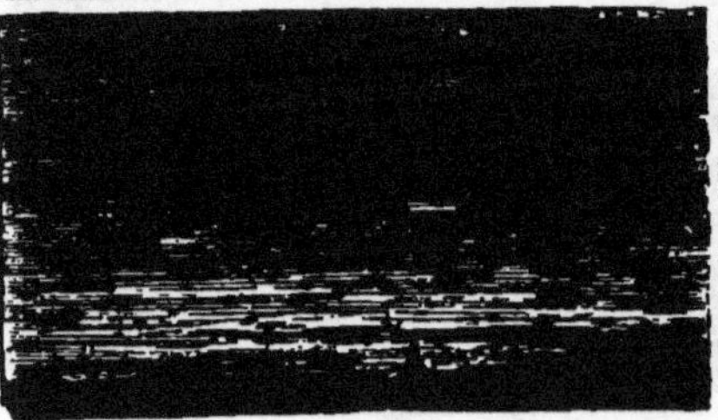

STRATUS.

The *stratus* cloud is a widely extended, continuous horizontal sheet, often covering the entire sky with a nearly uniform veil. This is the lowest of the clouds, and sometimes descends to the earth's surface. It is seen in the evening, and if it should disappear in the morning, the finest weather may be expected. When combined with the cirrus it forms the *cirro stratus* or mackerel sky, which indicates fair weather for that day, but rain a day or two after.

APRIL.

MOON'S PHASES.

New Moon ...	2nd	4.21 ev.	Full Moon....	17th	1. 4 mo.
First Quarter..	10th	10. 1 mo.	Third Quarter.	24th	3.40 mo.

		WEATHER PROVERBS AND WEATHER WISDOM.
1	MON	If it thunders on All Fool's day,
2	TUES	It brings good crops of corn and hay.
3	WED	If the first three days of April be foggy, there will be a
4	THU	flood in June.—*Huntingdon (Eng.) Proverb.*
5	FRI	A cold and moist April fills the cellar and fattens the
6	SAT	cow.—*Portuguese Proverb.*
7	**Sun**	**5th Sunday in Lent.**
8	MON	
9	TUES	SIGNS FROM THE RAINBOW.—If the green be large
10	WED	and bright in the rainbow, it is a sign of rain ; if the red be
11	THU	the strongest color, then there will be wind and rain
12	FRI	together. After a long drought the rainbow is a sign of rain ;
13	SAT	
14	**Sun**	**Sunday before Easter.**
15	MON	
16	TUES	after much wet it indicates fair weather. If it breaks up all at
17	WED	once, there will follow serene and settled weather. If the
18	THU	bow be seen in the morning, slight rain will follow ; if at
19	FRI	noon, settled and heavy rains ; if at night, fair weather.
20	SAT	The appearance of two or three rainbows indicate fair
21	**Sun**	**Easter Sunday.**
22	MON	weather for the present, but settled and heavy rains in two
23	TUES	or three days time.
24	WED	A severe autumn denotes a windy winter ; a windy
25	THU	winter, a rainy spring ; a rainy spring, a severe summer ;
26	FRI	a severe summer, a windy autumn, so that the air on a
27	SAT	balance is seldom debtor to itself.—*Lord Bacon.*
28	**Sun**	**Low Sunday.**
29	MON	Betwixt April and May, if there be rain,
30	TUES	It is worth more than oxen or grain.

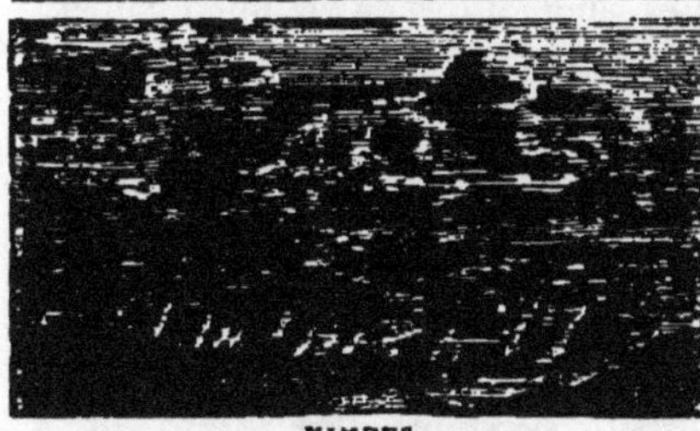

NIMBUS.

The *nimbus* is the true and immediate rain-cloud—shapeless but with defined outline, its edge gradually shaped off from the deep gray mass to transparency.

Just before rain we may observe what the sailors call "scud"— small under-clouds, often moving with much greater velocity than those above them, which seem sometimes stationary.

Sir John Herschel says "anvil shaped" clouds are likely to be followed by a gale of wind.

MOON'S PHASES.

New Moon...	2nd	7.57 mo.	Third Quarter..	23rd	9.48 ev.
First Quarter.	9th	5.39 ev.			
Full Moon....	16th	9.38 mo.	New Moon. ..	31st	8.54 ev.

WEATHER PROVERBS AND WEATHER WISDOM.

1	WED	Look at your corn in May,
2	THU	And you will come weeping away;
3	FRI	Look at the same in June,
4	SAT	And you'll come home in another tune
5	**Sun**	**2nd Sunday after Easter.**
6	MON	
7	TUES	A cold May enriches no one.
8	WED	A hot May makes a fat churchyard.
9	THU	A windy May makes a fair year.—*Portuguese Proverb.*
10	FRI	Water in May is bread all the year—*Spanish Proverb.*
11	SAT	
12	**Sun**	**3rd Sunday after Easter.**
13	MON	SUNRISE INDICATIONS.—
14	TUES	Above the rest, the sun who never lies,
15	WED	Foretells the change of weather in the skies,
16	THU	For if he rise unwilling to his race,
17	FRI	Clouds on his brow and spots upon his face;
18	SAT	
19	**Sun**	**4th Sunday after Easter.**
20	MON	Or if through mists he shoot his sullen beams,
21	TUES	Frugal of light, in loose and straggling streams,
22	WED	Suspect a drizzling day and southern rain,
23	THU	Fatal to fruits and flocks, and promised grain.—*Virgil, Georgic I, 438.*
24	FRI	
25	SAT	
26	**Sun**	**Rogation Sunday.**
27	MON	If red the sun begins his race,
28	TUES	Be sure the rain will fall apace,
29	WED	A high dawn indicates wind; a low dawn fine weather.
30	THU	A gray sky in the morning presages fine weather.
31	FRI	

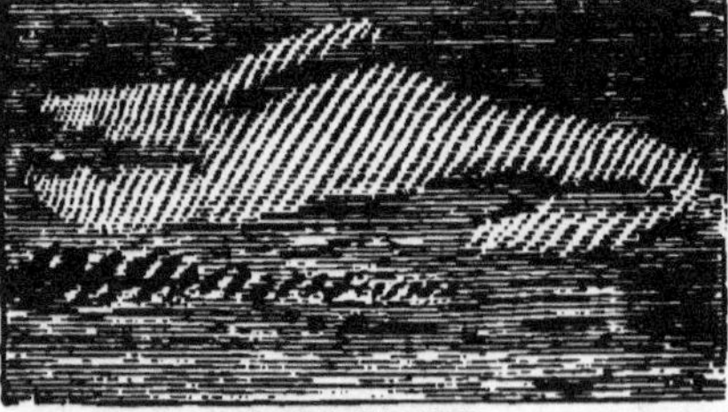

CIRRO STRATUS.

Cirro stratus clouds consist of delicate fibrous clouds spread out in strata, which are either horizontal or but slightly inclined to the horizon. Sometimes the whole sky is so mottled with this kind of cloud as to resemble the back of a mackerel, and it is hence called the *mackerel sky*. The cirro stratus invarably indicates wind and rain and is almost always to be seen in the intervals of storms.

John Macdonald & Co., Toronto.

STAPLE DEPARTMENT.

GREY AND WHITE COTTONS, PRINTS,
TICKS, DENIMS, FLANNELS,
BLANKETS, REPS, YARNS,
BATTINGS, WADDINGS,
TWINES, &c. &c,.

HENRY SKINNER,
Wholesale Druggist,

DEALER IN

SEEDS, PAINTS, OILS, DRUGS, DYE STUFFS, LAMPS, &c.

Coal Oil and Machinery Oils a specialty.

Correspondence solicited. KINGSTON, ONT.

W. TEES, Jr.,

Manufacturers' Agent, Commission Merchant, and Dealer in Plain and Fancy Furniture, 514 & 516 Craig Street.

Furniture

comprising Parlour Suites, Lounges, Easy Chairs, Chamber Setts, Sideboards, Dining and Centre Tables, Whatnots, &c., &c., at

LOW FIGURES.

made to order.

Bedsteads & Chairs in great variety.

A large assortment of Spring Beds, Mattresses, &c., &c., &c.,

Cheap for Cash.

Elegant Folding Chairs, Fancy Tables, Five o'clock Tea Tables, Brackets, Wall Pockets, Towel Racks, &c,, &c., SUITABLE FOR HOLIDAY GIFTS, all at prices to suit the times.

W. TEES, Jr., 514 & 516 Craig Street, Montreal.

"Send for our Annual Illustrated Circular."

60
FIRST PRIZES
AWARDED, AND
10,000
OF THOSE RAKES SOLD
IN FIVE YEARS.

Manufactured only by

G. M. COSSITT & BRO.,

Manufacturers of the BUCKEYE MOWER, ITHACA HORSE RAKE, REAPERS, SINGLE AND COMBINED SAWING MACHINES, SHINGLE MACHINES, 2, 4, 6, 8 and 10 Horse Power THRESHING MACHINES, &c., &c., &c.

Manufactory and Principal Office, BROCKVILLE, Ont.

COSSITT'S
Standard Implements.

Branch Office and Warerooms for the Province of Quebec :—

92 FOUNDLING STREET, MONTREAL.

R. J. LATIMER, Manager.

☞All Repairs kept, and Repairing done on our machines in connection with MONTREAL BRANCH.

THE OLD BUCKEYE

Old in name only, but new in every real

IMPROVEMENT

of the day. Small sections, short quick stroke, wrought iron and malleable used instead of cast, &c., &c., &c.

"Send for our Annual Illustrated Circular."

MOON'S PHASES.

First Quarter..	7th	11. 1 ev.	Third Quarter.	22nd	2.22 ev.
Full Moon....	14th	6.58 eve.	New Moon ...	30th	7.37 mo.

1	SAT
2	**Sun**
3	MON
4	TUES
5	WED
6	THU
7	FRI
8	SAT
9	**Sun**
10	MON
11	TUES
12	WED
13	THU
14	FRI
15	SAT
16	**Sun**
17	MON
18	TUES
19	WED
20	THU
21	FRI
22	SAT
23	**Sun**
24	MON
25	TUES
26	WED
27	THU
28	FRI
29	SAT
30	**Sun**

WEATHER PROVERBS AND WEATHER WISDOM.

A red sun has water in his eye.

Sunday after Ascension.

Fair weather for a week, with a southern wind, is likely to produce a drought if there has been much rain out of the south before.—*Admiral Fitzroy.*

If on the eighth of June it rain,
It foretells a wet harvest men sain.

Whit Sunday.

If it rain on June 8th (St. Medard) it will rain forty days later, but if it rain on June 19th (St. Protais), it rains for forty days after.—*French Proverb.*

If St. Vitus Day, (June 15th), be rainy weather,
It will rain for thirty days together.

Trinity Sunday.

A frequent change of wind with agitation in the clouds denotes a storm.

If the wind follows the sun's course expect fair weather.

When the wind veers against the sun,
Trust it not, for back t'will run.

1st Sunday after Trinity.

The southern wind
Doth play the trumpet to his purposes,
And by his hollow whistling in the leaves
Foretells a tempest and a blustering day.
—*Shakespeare, Henry IV.*

2nd Sunday after Trinity.

CIRRO CUMULUS.

Cirro cumulus clouds consist of small well defined rounded masses, in close proximity, and are generally formed by descending cirrus clouds. They are most frequent in summer, and on account of their fleecy appearance, they are sometimes called woolly clouds. When permanent they are a sign of increasing temperature and dry weather. In Buckinghamshire (Eng). they are called packets boys, and are said to be packets of rain soon to be opened.

JULY.

MOON'S PHASES.

First Quarter.	7th	3.27 mo.	Third Quarter..	22nd	7.22mo.
Full Moon....	14th	6. 1 mo.	New Moon . ..	29th	4.47 ev.

1	Mon
2	Tues
3	Wed
4	Thu
5	Fri
6	Sat
7	**Sun**
8	Mon
9	Tues
10	Wed
11	Thu
12	Fri
13	Sat
14	**Sun**
15	Mon
16	Tues
17	Wed
18	Thu
19	Fri
20	Sat
21	**Sun**
22	Mon
23	Tues
24	Wed
25	Thu
26	Fri
27	Sat
28	**Sun**
29	Mon
30	Tues
31	Wed

WEATHER PROVERBS AND WEATHER WISDOM.

If the first of July it be rainy weather,
It will drizzle for a month together.

If Bullion's day (July 4th) be dry, there will be a good harvest.

The moon with a circle brings water in herbeak.

3rd Sunday after Trinity.

The Aurora Borealis when very bright forebodes stormy, moist, unsettled weather.

A haze around the sun indicates rain ; it is caused by fine rain falling in the upper regions of the air ; when it occurs a rain of five or six hours duration may be expected.

4th Sunday after Trinity.

In this month is St. Swithin's day, (15)
On which, if that it rain they say,
Full forty days after it will
Of more or less some rain distill.

—*Poor Robin's Almanack, 1697.*

5th Sunday after Trinity.

A halo round the moon is an indication of rain, it being produced by fine rain in the upper regions of the atmosphere. The larger the halo the nearer the rain clouds and the sooner rain may be expected. A halo round the sun has often been followed by heavy rains.

6th Sunday after Trinity.

A shower of rain in July when the corn begins to fill
Is worth a plough of oxen and all belongs theretill.

CUMULO STRATUS.

Cumulo stratus clouds consist of the cumulus blended with the stratus, and are formed in the interval between the first appearance of the fleecy cumulus and the rain. On the approach of a thunder storm they are often seen in great magnificence, representing huge towers, rocks, and gigantic forms.

"When clouds appear like rocks and towers,
The earth's refreshed by frequent showers."

Manufacturers of MOWERS & REAPERS, HORSE HAY RAKES, THRESHING MACHINES, PLOUGHS, &c., &c., and all descriptions of FARMING IMPLEMENTS. Send for Price List.

THOUSANDS OF THE ABOVE PLOW IN USE.

LARMONTH & SONS, Agents, 33 College Street, Montreal.

THOMAS DAVIDSON & CO.,

JAPANNERS,

DOMINION STAMPING WORKS,

WHOLESALE TINWARE,

Factory : DOMINION, ALBERT & DELISLE STS.,
City Office : No. 27 WILLIAM STREET,

MONTREAL.

WILLIAM KENDALL,

Produce Shipper and Buyer

OF

Cheese, Butter, &c., &c.,

FOR THE ENGLISH MARKETS,

43 WILLIAM STREET, MONTREAL.

GOOD BOOK-KEEPING

To a man of business is equal to one-half of his capital.—*Mr. Commissioner Foubtangue, Court of Bankruptcy*, London, England.

DAY'S COMMERCIAL COLLEGE,

TORONTO,

(Established 1862.) A Select Business School for Young Men. Advantages offered : Individual and thorough instruction by an experienced Accountant, and course of study arranged to meet the capacity of pupils. For terms address, prepaid,

JAMES E. DAY, Accountant,

College Rooms, 96 King Street West.

JOHN MACDONALD & CO.,

keep constantly on hand one of the largest and most attractive stock of

Carpets and General House Furnishing Goods

in the Dominion.

NOVELTIES CONSTANTLY ARRIVING.

21, 23, 25, 27, Wellington Street,
30, 32, 34, 36, Front Street, TORONTO.

JOHN LAMB & SON,

ENGINEERS,

SPARKS ST., Ottawa.

The Construction of FLOURING, BARLEY, OATMEAL and SAW MILLS attended to.

A specialty made of our improved machinery for manufacturing Oatmeal, including our Patent Groat Screen for taking out Buckwheat, &c.

Gang Lath Machines, with Gangs of from two to ten saws.

Lamb's Patent Groat Screen, Wood Shapers, Band Saws, Derricks, Hoists, and all kinds of Grist and Saw Mill Furnishings made to order.

All sizes of LAMB'S DOUBLE ACTION WATER WHEEL kept on hand. The past record of this wheel proves it to be the BEST WHEEL now offered to the public.

Plans and Specifications of Mill Work, Surveys of Mill Sites, Machinery, Drawing, &c., made out at moderate rates.

A. CHRISTIE & CO.,

Chemists and Druggists,

SPARKS STREET, (corner METCALFE),

OTTAWA.

Country Physicians supplied with Medicines of same quality as those we use in dispensing.

Full supply of Trusses and Surgical Appliances.

Sheet Wax, White and of all colors, and all materials for the manufacture of Wax Flowers.

A. CHRISTIE & CO.

THE GREAT REMEDY.

WILSON'S

Pulmonary Cherry Balsam.

This perfect preparation, curing rapidly all Coughs and ordinary colds, is also a positive remedy for Bronchitis, Asthma, Quinsy, Laryngitis and Consumption.

It is a purely vegetable preparation, and by chemically combining the ingredients with great care and scientific skill we have a REMEDY THAT CAN BE RELIED ON. It quickly assimilates with the bone and blood, renewing the lung tissue, and making a marked improvement at once.

It enlivens the muscles, and assists the skin to perform its duties, and imparts strength to the system.

It loosens the phlegm, induces free spitting, and will be found very agreeable to take. It is not a violent remedy, but harmless in its nature—emollient, warming, searching and effective—powerful only in CURING ALL LUNG DISEASES.

A single trial will prove its efficacy in curing all Pulmonary Disorders over every other remedy known to mankind.

If all the Chemists in the country were to try and discover a specific for the cure of Throat and Lung Diseases, its CURATIVE POWERS COULD NOT BE GREATER THAN THOSE WHICH ARE WROUGHT BY WILSON'S PULMONARY CHERRY BALSAM.

This great remedy is performing too much good to make it necessary for us to do more than urge people to try it, and you will have no occasion to resort to other remedies, no matter how obstinate your cough may be.

As a remedy in Pulmonary diseases no medicine can obtain a higher or more deserved reputation.

It may be taken by old or young.

Sold in Bottles at 25 and 40 cents each—the 40 cent size containing double the 25 cent size. Sold by all druggists and dealers in medicine.

Address all orders to

J. W. BRAYLEY,

486 AND 488 ST. PAUL STREET, MONTREAL.

MOON'S PHASES.

First Quarter..	5th	8.26 mo.	Third Quarter.	20th	11.14 mo.
Full Moon....	12th	7.23 eve.	New Moon ...	28th	1. 6 mo.

		WEATHER PROVERBS AND WEATHER WISDOM.
1	THU	When it rains in August it rains honey and wine.
2	FRI	—*French and Spanish Proverbs.*
3	SAT	A wet August never brings dearth.—*Italian Proverb.*
4	**Sun**	**7th Sunday after Trinity.**
5	MON	When first the moon appears, if then she shrouds
6	TUES	Her silver crescent tipped with sable clouds,
7	WED	Conclude she bodes a tempest on the main,
8	THU	And brews for fields impetuous floods of rain ;
9	FRI	Or if her face with firey flushings glow,
10	SAT	Expect the rattling winds aloft to blow.
11	**Sun**	**8th Sunday after Trinity.**
12	MON	
13	TUES	But four nights old (for that's the surest sign),
14	WED	With sharpened horns, if glorious then she shine,
15	THU	Next day, not only that, but all the moon,
16	FRI	Till her revolving race be wholly run,
17	SAT	Are void of tempests both by land and sea.—*Virgil.*
18	**Sun**	**9th Sunday after Trinity.**
19	MON	
20	TUES	If it rain on St. Bartholomew's Day (24th August), it
21	WED	will rain forty days after.—*Roman Proverb.*
22	THU	
23	FRI	If the twenty-fourth of August be fair and clear,
24	SAT	Then hope for a prosperous autumn that year.
25	**Sun**	**10th Sunday after Trinity.**
26	MON	
27	TUES	For I fear a hurricane,
28	WED	Last night the moon had a golden rim,
29	THU	And to-night no moon I see.—*Longfellow, Wreck of*
30	FRI	*the Hesperus*
31	SAT	

BEST MODE OF OBSERVING CLOUDS.—In order to be able to distinguish well the form of clouds, it is often necessary to diminish their brilliancy by viewing them through a glass of a deep blue color, or by reflection from a mirror of black glass. We are thus able to detect peculiarities which entirely escape observation with the unassisted eye. —*Loomis.*

When cumulus clouds become heaped up to leeward during a strong wind at sunset, thunder may be expected during the night.

While any of the clouds, except the nimbus, retain their primitive forms, no rain can take place, and it is by observing the changes and transitions of cloud-form that weather may be predicted.—*Howard.*

MOON'S PHASES.

First Quarter.	3rd	3.32 eve.	Third Quarter..	19th	1.37 ev.
Full Moon....	11th	10.56 mo.	New Moon . ..	26th	9.17mo.

		WEATHER PROVERBS AND WEATHER WISDOM.
1	**Sun**	**11th Sunday after Trinity.**
2	Mon	
3	Tues	When it is evening, ye say it will be fair weather, for
4	Wed	the sky is red ; and in the morning, it will be foul weather
5	Thu	to-day, for the sky is red and lowering.—*Matthew*
6	Fri	*XVI.*, 2 and 3.
7	Sat	
8	**Sun**	**12th Sunday after Trinity.**
9	Mon	Holy Rood, August 14.—The passion flower blossomed
10	Tues	about this time ; the flower is said to present a resemblance
11	Wed	to the cross or rood, the nails, and the crown of thorns used
12	Thu	at the crucifixion.—*Circle of the seasons.*
13	Fri	A fruitful oak a long and hard winter.
14	Sat	
15	**Sun**	**13th Sunday after Trinity.**
16	Mon	If dry be the buck's horn,
17	Tues	On Holyrood morn,
18	Wed	'Tis worth a vest of gold ;
19	Thu	But if wet it be seen
20	Fri	E'er Holyrood e'en,
21	Sat	Bad harvest is foretold—*Yorkshire Provero.*
22	**Sun**	**14th Sunday after Trinity.**
23	Mon	A bright yellow sky at sunset presages wind ; a pale
24	Tues	yellow, wet--*Admiral Fitzroy.*
25	Wed	A dark gloomy blue sky is windy, but a light, bright
26	Thu	blue sky indicates fine weather ; when the sky is of a sickly
27	Fri	looking greenish hue, wind or rain may be expected—*Ibid.*
28	Sat	
29	**Sun**	**15th Sunday after Trinity.**
30	Mon	Dew and fog are indicators of fine weather.—*Ibid.*

Weather Indicated by Movements of Clouds.—If small clouds increase, expect much rain. If large clouds decrease, expect fair weather. Soft looking or delicate clouds foretell fine weather, with moderate or light breezes ; *hard edged* oily looking clouds, wind. As wind is only air in motion, its first effect is in diiving the clouds before it. Hence when clouds float about in a serene sky, from whatever quarter they come, you may expect wind. If they are collected in one place they will be dispersed by the rays of the sun. If they come from the north-east they indicate wind ; if from the south, great rains, but if, from whatever quarter, you see them driving about at sunset, they are sure signs of an approaching tempest.

OCTOBER.

MOON'S PHASES.

First Quarter..	3rd	2. 7 mo.	Third Quarter.	19th	2.16 mo.
Full Moon....	11th	4. 1 mo.	New Moon ...	25th	5.42 eve.

1 TUES
2 WED
3 THU
4 FRI
5 SAT
6 **Sun**
7 MON
8 TUES
9 WED
10 THU
11 FRI
12 SAT
13 **Sun**
14 MON
15 TUES
16 WED
17 THU
18 FRI
19 SAT
20 **Sun**
21 MON
22 TUES
23 WED
24 THU
25 FRI
26 SAT
27 **Sun**
28 MON
29 TUES
30 WED
31 THU

WEATHER PROVERBS AND WEATHER WISDOM.

If in the fall of the leaves in October many of them wither on the boughs and hang, it betokens a frosty winter and much snow.

16th Sunday after Trinity.

But more than all the setting sun survey,
When down the steps of heaven he drives the day ;
For oft we find him finishing his race,
With various colors erring on his face.
If fiery red his glowing globe descends,

17th Sunday after Trinity.

High winds and furious tempests he portends ;
But if his cheeks are swollen with livid blue,
He bodes wet weather by his watery hue ;
If dusky spots are varied on his brow,
And streaked with red a troubled color show,

18th Sunday after Trinity.

That sullen mixture shall at once declare
Winds, rain, and storms, and elemental war.
But if with purple rays he brings the light,
And a pure heaven resigns to quiet night,
No rising winds or falling storms are nigh—*Virgil.*

19th Sunday after Trinity.

If the sun sets behind a straight skirting of cloud, be sure of wind from the point where the sun is setting.

WEATHER INDICATED BY MOVEMENTS OF CLOUDS.—High upper clouds crossing the sun, moon, or stars, in a direction different from that of the lower clouds, or the wind then felt below, foretells a change of wind in that direction.—*Fitzroy.*

If two strata of clouds appear in hot weather to move in different directions, they indicate thunder ; if during dry weather, rain will follow.

A squall cloud that one can see through or under is not likely to bring, or be accompanied by so much wind as a dark continued cloud extending beyond the horizon.—*Fitzroy.*

Small scattering clouds, flying high in the south-west, foreshadow whirlwinds.—*Howard.*

NOVEMBER.

MOON'S PHASES.

First Quarter.	1st	4.57 eve.	Third Quarter..	17th	1. 5 ev.
Full Moon....	9th	9.40 eve.	New Moon . ..	24th	4.17mo.

1	FRI
2	SAT
3	**Sun**
4	MON
5	TUES
6	WED
7	THU
8	FRI
9	SAT
10	**Sun**
11	MON
12	TUES
13	WED
14	THU
15	FRI
16	SAT
17	**Sun**
18	MON
19	TUES
20	WED
21	THU
22	FRI
23	SAT
24	**Sun**
25	MON
26	TUES
27	WED
28	THU
29	FRI
30	SAT

WEATHER PROVERBS AND WEATHER WISDOM.

A warm and open winter portends a hot and dry summer.—*Lord Bacon.*

20th Sunday after Trinity.

Onions, skins very thin,
Mild winter coming in,
Onions, skins thick and tough,
Coming winter cold and rough.—*Gardener's Rhyme.*

An early winter, a surly winter.

21st Sunday after Trinity.

ST. MARTIN'S DAY, 11th.—If the wind is in the south-west at Martinmas, it keeps there till after Candlemas.

If cranes appear early in autumn, a serene winter is expected.

22nd Sunday after Trinity.

Much crying of peacocks denotes rain.

If dust whirls round in eddies when being blown about by the wind, it is a sign of rain.

Hares take to the open country before a snow storm. —*Scotch Proverb.*

23rd Sunday after Trinity.

When cats sneeze, it is a sign of rain.

If spaniels sleep more than usual, it foretells wet weather.

Bearded frost is a forerunner of snow.

He that would have a bad day, maun gang oot in a fog after a frost.—*Scotch Proverb.*

CLOUD INDICATIONS.—When the cirrus clouds appear at lower elevations than usual, and with a severe character, expect a storm from the opposite quarter to the clouds.

When after a clear frost long streaks of cirrus are seen with their ends bending towards each other as they recede from the zenith, and when they point to the north-east, a thaw and a south-west wind may be expected.

It is asserted that the versatile cloud pointed out by Hamlet to Polonias as "Very like a whale," was what is scientifically known as the cirro stratus.

If clouds look as if scratched by a hen (cirro stratus),
Get ready to reef your topsails then.—*Nautical Proverb.*

HOLMAN'S LIVER AND AGUE PAD

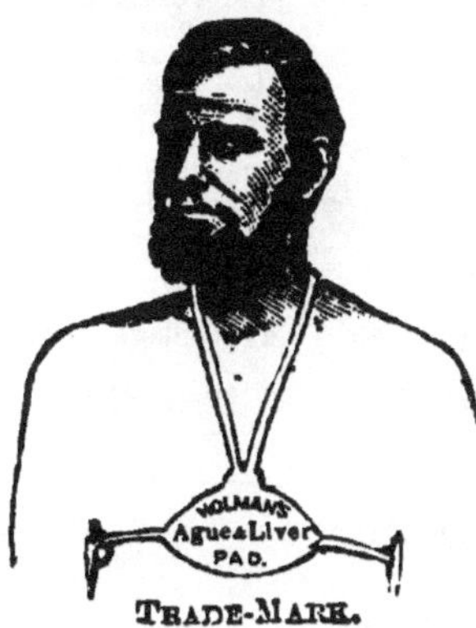

TRADE-MARK.

is marvellous in its prompt and radical cure for every species of Liver and Stomach difficulties. It contains only harmless vegetable compounds, and is worn EXACTLY WHERE NEEDED, over the vitals, the Liver and Stomach. It removes torpidity of the liver, and arouses the stomach from its dormant state, arresting fermentation, torpor and pain, by giving it the natural quantity of bile and gastric juice. It also vitalizes the entire system with Nature's true tonic. It arrests all deteriorated and poisonous fluids in the stomach, and thus prevents their entering the system by way of circulation. It absorbs from the body every particle of BLOOD POISON, whether bilious, malarial or medicinal, and leaves the wearer in perfect health.

Sent by mail, free of charge, on receipt of $2.50 for Regular Pad. Special Pads Three Dollars and a Half, used only in old complicated cases. Send money or Post Office Order, or Registered Letter, and you will receive the Pad by Mail, with book and full instructions.

Do not take any more medicine, but follow the example of thousands all over the world, by wearing one of HOLMAN'S PADS, a sure cure for Dyspepsia, Indigestion, Liver Complaint and all Bilious Disorders.

This is what the Pad is doing—Read the following testimonials :—

Holman Liver Pad Company, MONTREAL, Sept. 12th, 1877.

DEAR SIRS.—I am happy to relate to you the benefit I have received from your Pad, about one month ago. The pain I was troubled with is entirely gone, and my stomach is now in a perfectly healthy condition. The Pad has given me strength and removed the depressed feeling from which I was long suffering, thus making life a pleasure in place of a drag. My health was never better than it is at present, therefore, cheerfully I recommend HOLMAN'S LIVER PAD to suffering humanity, for it is worth more than its weight in gold.

Yours respectfully, JOHN J. HANNAN,
303 Notre Dame Street, Montreal.

Holman Liver Pad Company, BURLINGTON, ONT., July 30th, 1877.

DEAR SIRS.—With pleasure I communicate to you the benefit I have received in the use of your Fever and Ague Liver Pad. From the first day I put it on, the pain left me, and I now feel comfortable ; it also acted on my bowels like a charm, and I feel thankful to my Heavenly Father that my attention was directed to it, and also to you. I have certainly become your missionary for your Pad and Plasters.

Please find enclosed $3, for Pad and Plasters, and address them to Mrs. Hoggarth, Ingersoll, Ont.

Yours truly, JAMES C. BENT.

Holman Liver Pad Company, WYOMING, ONT., September 10th, 1877.

DEAR SIRS,—I have used your Pad for about three weeks, and feel great pleasure in informing you of its wonderful power and great benefit it has been to me, as I have been troubled with Dyspepsia for years. I wish your Pad gospel speed.

Yours truly, JOHN R. BENSON.

HOLMAN LIVER PAD COMPANY,

301 Notre Dame Street, Montreal, and 119 Hollis Street, Halifax, N.S.

LYMANS, CLARE & Co., *Wholesale Druggists*, AGENTS, MONTREAL.

Encourage Home Industry.

FYFE'S STANDARD SCALES have taken **SILVER MEDAL** and **FIRST PRIZE** for accuracy, workmanship and simplicity of construction. Hay Scales from $100, upwards. Platform and Counter Scales at low prices for good workmanship. RAILROAD TRACK, and DEPOT SCALES. Also, FANCY BRASS BEAM COUNTER SCALES.

Send for Illustrated Price List.

JAMES FYFE,
40 College Street
MONTREAL.

EVERY GENERAL STOREKEEPER

Visiting Toronto can fill his list in every department,

STAPLES, CARPETS, DRESS GOODS, HOSIERY, MANTLES, FLOWERS, WOOLLENS and NOTIONS,

- At -

JOHN MACDONALD & COMPANY.

IRA GOULD & SONS,

MILLERS,

MANUFACTURERS OF THE CELEBRATED BRANDS OF

"Gould's City Mills" Flour,

MONTREAL.

YOUNG & McGAURAN,

Boots and Shoes,

WHOLESALE,

28 COLLEGE STREET,

JAMES A. YOUNG,
(Late of Smardon & Young).
CHARLES McGAURAN.

MONTREAL.

MOON'S PHASES.

First Quarter..	1st	11.44 eve.	New Moon	23rd	4.31 eve.
Full Moon....	9th	2.56 eve.			
Third Quarter.	16th	10.10 eve.	First Quarter..	31st	9. 4 mo.

WEATHER PROVERBS AND WEATHER WISDOM.

1	**Sun**	**Advent Sunday.**
2	Mon	
3	Tues	Thunder in December presages fine weather.
4	Wed	If the sun shines through the apple tree on Christmas
5	Thu	day, there will be an abundant crop the following year.
6	Fri	A green Christmas makes a fat churchyard.
7	Sat	
8	**Sun**	**2nd Sunday in Advent.**
9	Mon	
10	Tues	If Christmas day on Thursday be,
11	Wed	A windy weather ye shall see ;
12	Thu	Windy weather in each week,
13	Fri	And hard tempests strong and thick ;
14	Sat	The summer shall be good and dry,
15	**Sun**	**3rd Sunday in Advent.**
16	Mon	Corn and beasts shall multiply ;
17	Tues	The year is good for lands to till,
18	Wed	Kings and princes shall die by skill.
19	Thu	Look at the weathercock on St. Thomas' day,
20	Fri	(December 21st) at twelve o'clock, and see which way the
21	Sat	wind is, for there it will stick for the next quarter.
22	**Sun**	**4th Sunday in Advent.**
23	Mon	
24	Tues	A windy Christmas and a calm Candlemas are signs of
25	Wed	a good year.
26	Thu	A warm Christmas, a cold Easter,
27	Fri	A green Christmas, a white Easter.—*German Proverb.*
28	Sat	
29	**Sun**	**1st Sunday after Christmas.**
30	Mon	If it rain much during the twelve days after Christmas,
31	Tues	it will be a wet year.

Cloud Indications.—And another storm brewing; I hear it sing i' the wind, yond same black cloud, yond huge one, looks like a foul bumbard that would shed his liquor * * yond' same cloud cannot chuse but fall by pailfuls.—*Shakespeare, Tempest.*

In summer or harvest, when the wind has been south for two or three days, and it grows very hot, and you see clouds rise with great white tops like towers, as if one were upon the top of another (cumulus), and joined together with black on the wether side, there will be thunder and rain suddenly. If two such clouds arise, one on either hand, it is time to make haste to shelter.- *Shepherd of Banbury*

J. R. JENNETT & CO.,

Importers of and Wholesale and Retail Dealers in

China, Glass, Earthenware, Electro-Plate, Cutlery, &c.,

Acadia Corner, 78 Upper Water Street,

1 and 3 Bells Lane, HALIFAX, N.S.

JOSEPH CARMAN,

General Commission Merchant,

WHOLESALE DEALER IN

Flour and other Produce, Provisions, &c., &c.

28 BEDFORD ROW, Halifax, N. S.

Goods Shipped Free of Cartage or Wharfage.

Consignments of Produce and Provisions solicited.

References : MONTREAL. Baird & Kinnear. GALT, Ont. R. Blain. TORONTO. D. Cowan & Co.

ADAM McKAY,

Manufacturer of Steam Engines, Boilers and Mill Machinery. Importer and Dealer in all kinds of Machinery and Machinery Supplies, Railway, Colliery and Contractors' Supplies,

Metal Merchant and Manufacturers' Agent.

Office and Sample Rooms: 30 SACKVILLE ST., HALIFAX.

Manufactory: DARTMOUTH, N. S.

YE

BOSTON BAKED BEANS,

In 3lb Cans, prepared in ye fashion of ye Olden Time in ye

ANCIENT CITY OF BOSTON,

By W. K. LEWIS & BROTHERS.

Ask ye Grocer for them.

Trade supplied by K. J. DOLPHIN, HALIFAX, N. S.

MEMORANDA.

THE YEAR 1877 REVIEWED.

JANUARY IN CANADA.

The first of January, 1877, found us well "snowed in," not only throughout the greater part of Canada, but also through the Northern United States, and as far south as Washington, D. C. The last week of the year 1876 was especially blustry, and the telegraphic reports of the weather from all quarters proclaimed "trains blockaded" and heavy drifts and snow falls. Altogether in December (1876) snow fell on 19 days, making a total snow fall of 23.6 inches, considerably more than the average for this month; while the rainfall registered 0. Thus, December of 1876 was a striking contrast to the December of 1875, when an unusual amount of rain fell, and but little snow.

MONTREAL RECORD.

In January, 1877, there were 14 clear brilliant days in Montreal, nine on which snow fell, three with rain or sleet, ten which might correctly be called *cold* days, and eight on which the weather was either extremely mild or thawing. The following is the daily record:

1. Bright, brilliant, mild, plenty of snow; great contrast to January, 1876. Country roads blocked up. Max. temp. 15° 9′; min. 4°.
2. Overcast, mild. Max. temp. 14° 8′; min. 6° 3′.
3. Brilliant, rather cold. 5° below zero during night.
4. Thermometer during night 5° below zero. Bright and clear.
5. Steady about zero. Brilliant weather.
6. Raw and overcast, and milder everywhere.
7. Heavy snow fall all day; evening mild.
8. Bright, mild day; great deal of snow everywhere. Sleighing as far south as Washington, D. C.
9. Thermometer during night 15° below zero. Bright cold day; thermometer steady at 10° below zero.
10. Cold day. Mild evening with sleet.
11. Very mild; wet snow in evening; hardly freezing.
12. Coldest night yet. 21° below zero. Very sudden and great descent of thermometer since yesterday.
13. Brilliant, cold—20° below zero.
14. Brilliant day. Third day of extreme cold.
15. Overcast; milder; spittings of snow; snowstorms to westward; snowed all night.
16. Great snow drifts everywhere, as far west as Chicago. Milder.
17. Bright, moderate day. Reports of blockades of snow from all parts.
18. Mild, overcast day.
19. do do rain in evening.
20. Wet night. Complete break up.
21. Overcast and moderate.
22. " " .
23. Spring-like weather. Brilliant.

24. Overcast and rather cold.
25. Brilliant and cold ; 10° below zero.
26. Mild, overcast day ; snowed heavily evening and night.
27. Snowing steadily during forenoon.
28. Mild day ; wet snow during night.
29. Very mild.
30. General thaw and break up. Snow fast disappearing.
31. Soft hazy morning ; thaw continues. Max. temp. 30° ; min. 25° 5'.

This record shows altogether a severe month. The greatest descent of the thermometer occurred between the eleventh and twelfth days, when the minimum readings, from 14° above, fell to 21° below zero within a very few hours.

In taking a general view of the month in Canada, we will dwell in no particular locality, but from Winnipeg to Halifax will gather and consolidate the general testimony which pronounces January, 1877, the severest in many years.

A candid correspondent in Toronto confesses on the 28th of December that "Vennor's last prediction is being fulfilled. A drifting snow-storm is raging here to-day. It has caused this morning's train from the east to be five hours late." On the night of the 29th of December, the snow cloud broke over Montreal, rendering pedestrianism very difficult, and on the same night Ottawa was clad in a snowy mantle nearly two feet in thickness. The hand which shed the snowflakes there did not withhold from other towns, but on December 30th, from Belleville, Toronto and Hamilton, arose a wail at the prodigality which sent snows sufficient to block up roads and delay trains. Again from Belleville, on January 1st, we hear that "the roads are badly drifted by last night's storm, but the stages were able to get through after much difficulty," and from Quebec, on the same date, that "a big snow storm set in during last night, and lasted with great violence until noon to-day. A very large amount of snow has fallen. The city streets are impassable, and communication with the country is cut off, consequently the New Year's markets to-day were completely bare. No mails have arrived from the west, and it is unknown when they will get in." A despatch from Halifax on the 2nd announced that "a snow storm prevailed all day," and two days later we heard that "snow-drifts still interfere with railway traffic." It was published in Montreal on the 6th that "the weather during the past few days has played sad havoc with water pipes in exposed situations, especially in dwellings where the foundation walls are supplied with ventilating chinks. The turbine wheel at the Water Works is totally incapacitated, being clogged up with ice until its revolutions have ceased." On the 8th we heard from Quebec that "the heaviest snow-storm of the season set in yesterday morning, and continued until the evening. About twelve inches of snow fell, and this is badly drifted. All roads blocked to-day." On the 9th, over one hundred men and about 70 horses and sleighs were at work in Montreal, in clearing the streets from the snowy encumbrances. Many of the narrow streets were almost impassable. From Ottawa and Kingston, despatches of the 11th announced that heavy snow-storms had set in, and the following day the thermometer fell to 12° below zero at Toronto, and at Rockliffe, on the Ottawa, registered 34° below zero. At London it snowed all day on the 12th, and a Kingston despatch of the same date says: "Last night there was a blinding snow drift with wind from the north-east. This morning it was very cold, the thermometer being down as low as 20 degrees below zero." At Quebec the thermometer at the Cape on the same day stood 30 degrees below zero, and a Halifax despatch says, "snow is deposited six feet deep in the woods, and lumbering operations are being vigorously prosecuted. Another cold spell has set in." From Manitoba it was learned that "the thermometer on the 11th, at Winnipeg, stood 40 degrees below zero; on the 10th, at Swan River, 47 degs., and at Battleford, 52 degs. below zero." On the 13th the thermometer showed 16 degrees below zero at Guelph. A Toronto dispatch of the 14th says "Grand Trunk trains east and west have failed to arrive here, on account of the heavy drifts on the line;" in fact in all directions, great inconvenience was experienced during this week by railway companies. The train from New York on the 14th occupied 48 hours in the trip to Montreal. A heavy snow-storm visited Quebec on the 14th, and travelling westward, its precursor, a heavy gale, swept over Montreal on the 15th, bringing with it the snow-storm in the evening. The herald of the storm was not long reaching Toronto, and on the 16th the train from Belleville was detained for 25 hours near Bowmanville, some of the passengers finding it difficult to obtain food. In one instance three engines were coupled to a train, but this proved ineffectual against the huge snow banks. Even St. Catharines and London felt the effects of this storm, as a despatch from St. Catharines of the 16th says: "Traffic on the Welland Railroad was almost suspended yesterday owing to the snow blockade. * * * The deep cutting between this city and the port was entirely blocked up, and no train was able to get through until about 11 o'clock this morning, when they succeeded in forcing their way through with two engines." On the 15th a despatch from Halifax stated that the mercury stood 10 degrees below zero, the severest cold felt up to that date. On the same day twelve inches of snow fell at Guelph, which very quickly drifted, and at St. John, N.B., there was a heavy storm, which delayed all railway operations. From Ottawa, on the 17th, it was despatched that "the exceptionally heavy storm of last night had the effect of temporarily checking traffic along the line of the Grand Trunk. The Montreal delegation to the Dominion Board of Trade reached the capital at noon, but the Toronto and western members were snowed up at Scarboro, where, from all that has been learned, they remain." From Kingston, of the same date we learned that "the roads in this vicinity are completely blocked with snow. The eastern and western mails were very late; no American mails." At Halifax snow fell on the same date, blocking up the country roads. On the 18th a despatch from Winnipeg stated that "the thermometer registered 44 degrees below zero at Winnipeg, at Pembina 47, and at Fort Pelly 52 degrees below zero." On the 23rd, intelligence from St. John, N. B., stated that "the thermometer was reported as low as 39 and 40 degrees below zero at Andover and other parts of Victoria County."

Perhaps the most singular event in the weather of the month was the occurrence of a thunderstorm on the 20th, an account of which was received from Quebec on the 24th, as follows: "A heavy thunder-storm visited the parishes between Rivière du Loup and Rimouski on Saturday night. At Cacouna the Roman Catholic church steeple was struck by lightning and set on fire, but the flames were extinguished before much damage was done." On the 26th the thermometer registered 12 degrees below zero at Halifax. Two days later a heavy snow-storm occurred at Ottawa, and on the 30th a despatch from Quebec says: "A terrific snow-storm set in yesterday, which continued during the night. The drift is deep, and some of the country roads are impassable."

With such a record as the foregoing, no honest student of the weather can do otherwise than admit that the January of 1877 was unusually severe and wintry-like; yet anti-Vennorites, hugging the mild and exceptional February which followed it, ignored three repeated snow-falls and severe snaps, and towards spring talked loudly of the "little snow" and "little cold" which had visited us during the winter months. The facts of the case, however, are that not for fifty years has there been a January during which the cold and snow extended so far east, west and south as during that of 1877. The winter thus having pretty well spent itself during January, February came in smiling and pleasant, and brought with it the thaw expected some days before.

JANUARY IN THE UNITED STATES.

Correspondence from a very wide area, together with information gleaned from newspapers and various other sources, will serve to review the general character of the month throughout the United States. We hear from Boston of December in Massachusetts that "the snow lay from two and a half to three feet on the level; ice was 16 inches thick. Snowy owls abounded in such numbers that a taxidermist had 60 on hand at one time to stuff."

The old year did not expire without a struggle, but from east and west we hear of frost and storm. On the 28th of December a snow-storm destroyed nine houses near Alta, and buried the occupants, the news of the catastrophe being dispatched from Salt Lake four days later.

The Washington correspondent to the Montreal *Gazette* writes on January 3rd that "the innocent little New Year was ushered into this tempestuous world on the wings of a regular northeast snow-storm, which kept increasing during the whole day, till at nightfall the streets were nearly deserted. The snow and wind had the out-of-door city to themselves, and right merrily did they dance and whirl and sing, holding grand New Year's carnival; or was it a wake at the funeral of the poor old careworn century just departed that they were celebrating so boisterously?—and was this large white mantle, encircling everything around, his winding sheet." Again, on the 6th we hear from Washington that the continued cold weather, followed on New Years Day by the most severe snow-storm experienced within ten years, had driven game south, and on January 8th there appeared in the *Witness* the startling announcement that the estimated loss to New York city by the great snow-storm was $1,000,000. Even in Texas the snow lay several inches deep for some consecutive days during the month of December, and in Tampa snow fell on New Years Day, a phenomenon never witnessed before. The *Florida New Yorker*, referring to the weather experienced in Florida during the early part of January, 1877, says: "It has been a matter of nearly a month's anxious enquiry to learn through the Florida papers, as well as diligent private correspondence, the exact amount of injury sustained by this *unprecedented spell of weather*. Beyond the killing of very young trees in certain sections of the State, partial injury to fruit, and the shedding of the foliage, no very serious loss has been sustained. To be able to make this report would be worth thousands to Florida. It may be fifty years before such an extraordinary weather report shall be again registered." Prof. Brown Goode, of the Smithsonian Institute, writes from Bermuda that the same cold weather had been felt there which the northerners had experienced, but that the thermometer had at no time registered lower than 54 degrees above zero. The *Witness* of January 8th says: "Snow in plenty has been falling in the United States, as well as in Canada, fulfilling Mr. Vennor's predictions that there would be a great deal of it this month. Different cities give it different receptions. Washington is rejoiced, and turns out in sleighs filled with masqueraders, but commercial New York complains of the snow blockade and the interruption of business, while Buffalo, deeply interested in transportation, sends a message about delayed trains."

From Connersville, Indiana, on the 10th, it was stated that for three weeks heavy snows and extremely cold weather had been the order of the day, and on the same date a despatch from Washington says: "The heavy fall of snow on New Years day has given Washingtonians such a taste of real winter and its attendant comforts and discomforts as they have not experienced for the last fifty years. The rain on the 6th brought the festive part to a close, and was celebrated by a sleighing carnival on Pennsylvania Avenue." On the 13th, *Forest and Stream* reported that the weather had been very severe in the Adirondacks, in fact the hardest winter known in that latitude for years, the snow being deep and the cold intense.

Extracts might be multiplied to fill page after page, but a sufficiently clear idea of January weather will be formed from the foregoing.

FEBRUARY.

The month of February, 1877, will be a memorable one for its exceeding mildness and the small amount of moisture which fell from the clouds, either in the shape of rain or snow. The soft weather which set in after the 18th of January continued almost without interruption up to the 12th of this month, when for a brief space everything was firmly frozen again. After the 21st balmy, spring-like weather set in, the snow rapidly disappeared, and waggons and carriages appeared in the streets. The last of the month was bright and balmy. There were 15 brilliant days at Montreal this month out of the 28, and only some three or four that were at all cold. Snow fell on six and rain on four days. The minimum temperature of the three coldest days in the month at Montreal was : 13th, 2° 2′ ; 14th, 5° 2′; 19th, 5° 6′. During eight days the lowest reading of the thermometer fluctuated about the freezing point only, while in two days it was above.

On the 12th of the month a despatch from Toronto, concerning the weather, states that "Vennor seems to have come out right at last, if he is allowed a week or ten days' time. Old Boreas came down to-day good and strong, and everything is freezing up. Ulsters were very conspicuous in the streets this afternoon, the breeze from the north being very inviting." A despatch from St. John, N.B., to the Montreal *Herald*, dated the 15th, confirms the reports of heavy snows, declaring that "trains on all parts of the Western Extension and local railways have been, and are still, greatly delayed by yesterday's storm ; on the Western road snow ploughs have gone through drifts eight feet deep." On the same date it was learned from Ormstown, Q., that "the mild, soft weather of the past few weeks was checked by a snow flurry on Monday evening, since which time the weather has been fair and frosty. Further east considerable snow has apparently fallen, in many places blocking up the roads. There are at present meteorological indications of a snow-storm before long—very likely between now and the 20th inst." The Montreal *Witness* of the 19th, referring to the extensive snows in February, says : "Two letters received on Saturday, one from Rochester and another from twenty miles west of this place, speak of unusual quantities of snow. The tops of the fences are covered along many miles of road, and there are drifts the like of which have not been seen for many years. On the 13th there was also a heavy blockade of snow at St. John, N.B. So far we have escaped these snow-falls, and have enjoyed the 'brilliant' weather anticipated by Mr. Vennor, whilst the other parts of the country just referred to have come in almost to date for his snow-storms." From Douglastown, Gaspé, it was despatched on the 24th : "There have been heavy snow-falls here since the 17th of the month, and the snow is from seven to eight feet deep in many places. Travel was completely stopped for some days, and has been until recently much impeded. We have not had so much snow here during February for years."

Among a popular school of cynics there has been a disposition to cavil at the sometimes imperfect realization of Vennor's probabilities, but even in this month, which was characterized by such marked peculiarities, the general tenor of his predictions was proved to be correct.

RECORD AT MONTREAL.

		Min. Temp.
1.	Thaw continues ; slush ; overcast ; rain in evening....	27° 8′
2.	Overcast ; dull ; slushy..........................	33° 8′
3.	Thaw continues ; slight snow......................	31° 8′
4.	Mild ; alternate cloud and sunshine.................	24° 5′

5.	Hard frost ; thaw towards evening	20° 4′
6.	Thawing ; great slush	
7.	Thaw continues ; light snow ; wind in evening	28° 0′
8.	Brilliant day ; thawing only in sun ; cold night	13° 1′
9.	Brilliant frosty day, but sun powerful	13° 5′
10.	Overcast all day	13° 7′
11.	Brilliant day ; colder	18° 2′
12.	Dull ; rain and sleet ; flurries of snow ; snow fall all night.	
13.	Brilliant and cold ; snow-storms at points west	2° 2′
14.	Brilliant cold day	5° 2′
15.	Bright and thawing	14° 6′
16.	Mild ; snowing ; rain ; cloudy	28° 9′
17.	Brilliant and cold	12° 7′
18.	Brilliant cold morning ; cutting wind ; snow-storm north of Toronto	11° 3′
19.	Brilliant cold day	5° 6′
20.	Brilliant and cold ; snow in afternoon	17° 0′
21.	Brilliant, balmy weather	27° 0′
22.	Brilliant, spring-like weather ; carts and carriages out	28° 7′
23.	" " snow nearly all gone	25° 7′
24.	Cloudy, raw morning ; rain and sleet in evening ; snow at night	29° 6′
25.	Mild ; light snow in the morning	30° 0′
26.	Cloudy day ; gales in New York and Long Island	29° 0′
27.	Brilliant day ; fairly cold ; sleighing wretched	18° 2′
28.	Brilliant, balmy, spring-like day	13° 0′

MARCH.

It may be well in reviewing the weather of this month to glance at several of its prominent features previous to a minute retrospect from different localities. About the 8th and 9th heavy gales swept over Canada and the United States. St. Patrick's day was cold, with snow in Halifax and sleet in Washington, D.C., and the month ended, at most places, quietly, with early signs of spring, and the opening of navigation as predicted. The gales throughout the United States were furious, and telegraphic despatches from various places report very serious damage. From Boston, on the 10th, it was telegraphed that "the gale was the severest in this vicinity for years ; the velocity of the wind at 10.30 a.m. to-day was 72 miles an hour. Much damage to buildings is reported." The wind had not shown such velocity since the establishment of a signal station there. From New York, of the same date, a despatch says : "The roof of the German Catholic church, 125th street, was blown off last night. Nothing has been heard of the steamship 'Amerique ;' the wires are down, and there is too heavy a sea for the Long Branch boat. A house was blown down at Elizabeth, and the lower part of that city was damaged by the high tide. In Brooklyn a large four story frame building belonging to the Chemical Works was demolished." Another despatch of the 10th, from Taunton, says: "The gale here was very severe, blowing down chimneys, trees, signs and fences. The greater part of the roofing of the extension of Leonard block, built for a theatre, was lifted by the wind, carried a long distance, like a huge balloon in the air, over a block of buildings, and dumped in the main street, carrying away a lamp post and frightening the occupants of stores."

The snow and cold of the month are telegraphed from all sources. On the 2nd a dispatch from Quebec said: "A big snow-storm set in from the east this afternoon, and still prevails with considerable fury," and on the 8th another storm visited Quebec, and a great quantity of snow fell. On the same date, from St. Louis, a despatch reads: "The most violent snow-storm of the winter prevailed here to-day; travel is much impeded. The storm originated in Colorado, and has extended over Kansas, Indian Territory, part of Arkansas and Missouri, and is now travelling eastwardly and north-eastwardly." A snow-storm set in at Montreal on the 9th, and on the 12th a punster writes from Cartwright that "The weather has a very Vennor-able appearance just now. Good sleighing, for which those having ties and wood to get to the front are truly thankful." On the 13th a despatch from Salt Lake City says "Snow has been falling in the mountains here for twelve days," and the next day an Ottawa despatch said "plenty of snow in the bush is reported from up the river, and the drawing of timber is progressing favorably." Another despatch of the same date from Sarnia says that "the street railway has been blockaded since the big snow-storm. A force was set to work on Saturday to clean it off, but the snow-storms of Saturday and Monday neutralized their labors. The track was cleared again yesterday, and travel was resumed last evening." Another correspondent, of the same place, writes on the 14th: "Vennor wasn't very far wrong about that three feet of snow in March, was he? If it was a guess it was a very lucky one, and has done much to re-establish his reputation, somewhat damaged by his failure to bring on those February storms according to programme;" and the Sarnia *Canadian* of the same date says: "People who grumbled because February was unseasonably fine have had no reason to complain of March on that score. Last Thursday ushered in a real old fashioned snow-storm, accompanied by a gale of unusual violence, and next morning found the snow some twelve inches deep on the level. Sunday brought on another snow-storm, and Monday still another of more than common violence, though, fortunately, not accompanied by cold weather." On the 16th a despatch from Peterboro said: "Although we have not as yet got the quantity of snow predicted by Mr. Vennor, yet sufficient has come to give us sleighing once more, and advantage is being taken of it by the farmers in particular in completing their contracts for delivery of wood, grain, etc."

On the 19th, trains east and west were delayed by blockades near St. John, N.B., and on the 20th news from Stratford reported a heavy fall of snow, accompanied by thunder and lightning, and the same day a heavy snow-storm raged at Chicago, impeding travel greatly. "Wednesday morning, the 21st, saw a small army of villagers, in Elora, Ont., clearing away from their approaches the heavy fall of snow of the previous evening." The despatch continues: "We haven't quite the three feet yet, but rapid strides are being made in that direction." The same day a message from Halifax says "it has been raining here for forty-eight hours." A St. Louis despatch of the 24th states, "the severest snow-storm of the season occurred yesterday," and from Toronto on the 30th it was telegraphed: "The snow blockade still continues on the Toronto and Nipissing RR., and no trains left here to-day."

The early signs of spring and opening of navigation were apparent in most places, as a few extracts will show. A Toronto despatch of the 9th affirms: "The ice disappeared from the bay at Toronto on the morning of Saturday last, and soon after daylight a vessel was under sail, and placed in position to receive a cargo of peas;" but on the 13th "the bay was covered with a coating of ice an inch thick." From Albany, on the 27th, a message arrived: "The ice barrier below the city having disappeared, the river is now open to New York. A steamboat reached the city to-day from Coeymans." The Montreal *Witness* of the 27th says: "Wild geese are again making their appearance, going west, which some people look upon as an indication of an early spring. The ice-bridge opposite the city is becoming flooded with water, owing to the rain of last night and to-day, which is likely to bring crossing by sleighs to a close earlier than was expected. Very little snow is to be seen in the country now, and the roads are almost bare." By Toronto despatch of the same date we learn that "The ice went out of the bay yesterday with a rush; several ice-boats which were flitting to and fro, notwithstanding that it was Sunday, were suddenly immersed in water, and considerable difficulty was experienced in landing them and their occupants." A day later intelligence from Halifax says: "The wet weather continues. Nearly all the harbors on the eastern coast and in Cape Breton are now open."

MONTREAL RECORD.

1. Brilliant day; aurora visible at night.
2. Overcast and fairly cold; snow, sleet, rain and wind—March bluster; snow-storm at Ottawa and Quebec; thunder-storm at Hamilton.
3. Windy and overcast.
4. Snowed during night; light rain; colder towards afternoon, and clear.
5. Bright, cold morning; snow in the afternoon; mean temp. 20°.
6. Bright, cold morning; mean temperature 14°.
7. Six inches of snow last night; bright, cold, drifty morning; more like winter again; cold, windy night; mean temperature 20°.
8. Snowing and drifting all day; great snow-storm at Three Rivers; rain at night; mean temperature 12° 9′.
9. Rain and great slush in the morning; at noon hurricane of wind.
10. Cold, bright day.
11. Raw and cold; fitful sunshine; snow in evening.
12. Bright and cloudy day; mild.
13. Bright, cold morning; thermometer at 10°.
14. Bright, cold morning; overcast at noon; snow at 6 p.m.

15. Five or six inches snow fell during night; therm. at 9 p.m. 15°.
16. Bright morning; cloudy afternoon, with light snow.
17. *St. Patrick's Day.* Brilliant, cold day; 4° below zero during night; 18° below reported at Fort Garry; snow at Halifax all day; snow and sleet at Washington, D.C.
18. Very cold last night; cutting N.E. wind; thermometer 7° below zero during the night; snow-storm at St. Johns, N.B.; mean temperature 5°.
19. Brilliant day; milder.
20. Brilliant morning, with cold wind.
21. Snowing hard at Montreal, Stratford and Chicago.
22. Bright, mild, spring-like day; great slush.
23. Overcast day, with falling barometer.
24. Overcast; spring-like; raining forty-eight hours at Halifax.
25. Overcast day; raw northerly wind; rain in evening.
26. Overcast and mild; great snow-storm at St. Louis; rain at Montreal at 6 p.m.; navigation opening at Toronto; early spring certain.
27. Rain all last night and this morning; raining yesterday at Toronto, Guelph, Stratford, Belleville, Kingston, Brockville, and many other points west; navigation open at Toronto Bay, Albany, etc., early as predicted; snow nearly all gone; river ice still firm at Montreal; robins seen.
28. Rain all night and this morning; heavy east gale and rain-storm at Quebec and Halifax; great rise in rivers; ice at Montreal flooded; few crossing; rained all day.
29. Snowing in forenoon; ice giving way at Quebec; rains reported everywhere; snowing at Belleville, Kingston, Brockville and Cornwall yesterday; snowing heavily at 9 p.m. at Montreal.
30. *Good Friday.* Bright day; sun warm and spring-like; very little snow or ice left.
31. Brilliant, summer-like day; March ends lamb-like as predicted; temperature—maximum, 46°, minimum, 30° 6'.

APRIL.

The months of the year 1877 seem to have been somewhat misarranged, or rather to have interchanged characteristics so frequently as to have almost destroyed their identity. April wielded the genial influences of May, and developed vegetation to an unusual degree, while the clear, dry weather which prevailed robbed Spring of the dampness which usually precedes Nature's resurrection. As with previous months, we will introduce a few extracts from telegraphic reports, showing the general character of the weather. On the 2nd a despatch from Winnipeg reads: "Several days of mild weather caused the roads to break up; ice on the rivers is getting rotten. Hawks were seen two weeks earlier than migratory birds usually appear." Intelligence of heavy floods, caused by the breaking up of the rivers, was received on the 4th from St. John, N.B., reporting serious loss of life at Gaspareaux, where several mill dams were swept away. Showing the early opening of navigation, a telegram from Kingston on the 5th said: "This afternoon the steamer will make a start for the Island and Cape Vincent." On the 6th the bridge opposite Montreal "caved in near both shores, and a general shove took place," and a despatch from Halifax on the 9th says, "Pugwash harbor is open and no ice is visible."

On the 23rd, a despatch from Winnipeg says "the weather is warm. * * Navigation is expected to be open in a few days." On the 25th an Ottawa telegram reports that "a little boy was affected by sunstroke on Monday (23rd) while playing on York street. The thermometer registered 73° in the shade;" and a Quebec despatch of the same date says: "The warm weather of the past few days ended this morning in a thunderstorm. It is now cool and cloudy, with indications of rain." From Quebec, on the 28th, we learn: "Farmers from the surrounding country parishes report that the snow has nearly disappeared, and crops are being rapidly put in. They are fully three weeks in advance of last year."

WEATHER RECORD AT MONTREAL.

1. Overcast day; light rain during early part of night; maximum temperature 46 deg. 9m.
2. Overcast, balmy day; max. temp. 48 deg. 1m.
3. Cold night; brilliant morning; keen N.E. wind; river ice very shaky; dust on roads; hardly a trace of snow left; max. temp. 35 deg.
4. Bright, cold morning; froze hard last night; roads in many places dry and dusty; river open above the Nuns' Island; road to Laprairie impassable; max. tem. 43 deg.
5. Bright, warm morning; navigation open at Kingston, and boat running; ice still firm opposite Montreal; dusty roads; robins and song sparrows arrived; max. temp. 44 deg. 4m.
6. Overcast morning, with spittings of snow and rain; rained briskly towards 5 p.m.; day raw and cold; ice broken along both shores of river; rained through part of night; max. tem. 47°.
7. Bright, cloudless day; cold north wind; sun powerful; max. temp. 43 deg.
8. Brilliant summer-like day; easterly wind; max. temp. 47 deg. 7m.
9. Bright, warm, dusty, summer-like day; ice broken up all down channel of river; steamboat running on Richelieu to Belœil; very advanced season; max. temp. 54 deg. 7m.
10. Warm, summer-like day; street cars commenced running; great ice shove—ice two feet thick; max. temp. 55 deg. 6m.
11. Summer weather; very dusty; river channel well open; no swallows yet; max. temp. 56 deg. 4m.
12. Brilliant day; cold N.E. wind; dust in clouds; ice not yet left front of city, but open everywhere else; maximum temp. 49 deg. 1m.
13. Hazy morning; calm and bright; warm and very dusty; ice still jammed up in river; max. temp. 55 deg. 7m.
14. Same as last three days; dusty, warm summer weather; very dry period; max. temp. 58 deg.

15. Calm, balmy weather ; no signs of rain ; ice left front of city last night ; max. temp. 61 deg.
16. Cloudy day ; dust fearful ; water below wharves ; ice, as a body, gone ; open for boats ; max. temp. 52 deg.
17. Cloudy day ; first steamboat arrived—10 days earlier than last season ; rain at 2 p.m. ; tug W. C. Francis first vessel in, shortly followed by others ; max. temp. 57 deg.
18. Rained a little last night ; warm, bright morning ; vegetation advancing rapidly ; more like May than April ; several steamboats and other river craft in ; max. temp. 56 deg. 8m.
19. Rained all night ; sleet and snow in morning ; cold rain all day ; no swallows yet ; max. temp. 46 deg.
20. Rain, snow, sleet ; ground this morning white ; all slush ; cold rain all day ; max. temp. 39 deg.
21. Ground white with snow this morning ; snowing briskly at 7, 8 and 9 a.m. ; snow in London, Eng. ; wintry day ; maximum temp. 41 deg. 6m.
22. Warm, summer-like day ; swallows arrived in considerable numbers ; max. temp. 65 deg.
23. Brilliant, summer-like weather ; swallows in every direction, and other birds ; max. temp. 73 deg.
24. Hot day—unusual warmth for the season ; shower at night ; max. temp. 74 deg. 2m.
25. Much cooler ; thunderstorm at Quebec yesterday ; weather cool and pleasant all day ; max. temp. 61 deg. 3m.
26. Beautiful summer weather ; snow to the far West yesterday ; max. temp. 61 deg.
27. Dry, summer-like weather ; max. temp. 61 deg. 3m.
28. Cloudy day ; rained evening and all night ; max. temp. 55° 4′.
29. Raining this morning ; afternoon fine ; evening and night again wet ; max. temp. 61 deg. 2m.
30. Raining hard ; cold rain ; cloudy evening and night ; maximum temp. 54 deg. 2m.

MAY.

May's genial characteristics were not lacking in 1877, for early in the month the Toronto *Mail* says : "It is the general talk that not for years have we had so favorable a spring for farmers as the present one." An Ottawa telegram of the 9th says : "Farmers report that the crops are pretty well in. The weather has been favorable, and all that is now wanted to secure a bountiful harvest is occasional showers of rain." We may here be permitted to advert for a moment to a letter from France which appeared in the Montreal *Witness* of the 9th, in which the correspondent writes : "Spring continues to be as capricious as a woman—one day a smile and the next a concealment, mocking poor citizens." But to return to the weather in Canada. An exception to the general fine weather was felt in New Brunswick, as a despatch of the

10th says: "A foot of snow fell on Tuesday over the country between Sussex and Sackville, and much snow in the Gulf of St. Lawrence district." Two days later a telegram from Quebec says "agricultural reports from the surrounding country districts are very gratifying." On the 15th, despatches from Halifax, Ottawa, Boston, etc., report great bush fires, the smoke of which, like dense clouds, hung over the country, so that in many places lamps had to be lit much earlier in the evening than usual. Intelligence of very extensive bush fires continued to arrive on the 16th, and a telegram from Winnipeg reports "Weather warm; two cases of sunstroke; crops all in." A weather item in the *Witness* on the 17th stated that "the long expected rain which fell last night was much needed, nearly three weeks having elapsed since the last rain fall. Gardens and trees look very much improved this morning by it." On the 18th a very disastrous storm, which blew down the Roman Catholic Church at St. Hypolite, was telegraphed from Joliette and St. Jerome. The Rev. Mr. Boileau, Vicar of St. Jerome, was instantly killed, and his son fatally injured. The New York *World*, commenting on the weather on the 18th, says: "We do not often talk about the weather, but the weather of yesterday forces itself into discussion. The heat was almost unprecedented so early in the season;" and the *Witness* of the 25th, referring to it, says: "Vegetation is at present extraordinarily advanced, and there have not been two Queen's Birthdays in a score of years which have been ushered in with such luxurious foliage as was our 24th this year."

MONTREAL RECORD.

1. Cloudy, cold morning—cold enough for snow; snow in Sarnia; max. temp. 59 deg. 3m.
2. Cold, cloudy morning; flurry of snow and sleet towards evening; very cold night for the season; max. temp. 44 deg. 7m.
3. Cold morning; overcast afternoon; snow-storm at Three Rivers; max. temp. 48 deg. 2m; min. 32 deg. 2m.
4. Beautiful cloudless day; warmer; severe frost last night; max. temp. 54 deg. 2m; min. 36 deg. 6m.
5. Brilliant, warm summer morning; fountains commenced to play in public gardens and squares; max. temp. 56 deg. 2m.
6. Warm, summer-like day; wind cool; max. temp. 61 deg. 3m.
7. Warm, summer-like day; clouded over in afternoon; rain towards 5 p.m.—light shower; vegetation unusually advanced, and advancing most rapidly; max. temp. 51 deg. 7m
8. Cold again, and cloudy; snow in lower provinces.
9. Cold, windy, fall-like day; stormy night; max. temp. 49 deg. 9m.
10. Warmer; wind and great dust; cloudy; rain much needed.
11. Bright, beautiful day; wind cool.
12. Warm, hazy day; very summer-like—more like June day; hot afternoon; ice-cream and soda-water going; vegetation wonderfully advanced for the season; max. temp. 68 deg.
13. Summer day—*hot;* air filled with smoke from fires; very dry spell; max. temp. 72 deg. 5m.
14. Gale of wind last night; wind and rain at 5 p.m.; air full of smoke.

15. Warm weather; bush fires everywhere—dense smoke; rain much required; max. temp. 74 deg. 9m.
16. Severe thunderstorm towards 10 p.m., with heavy rain.
17. Close, sultry day; thunder clouds; more like June or July weather than May.
18. Hot, sultry weather; intensely hot this morning; heavy showers during afternoon, with thunder.
19. Bright, hot morning; young robins old enough to fly.
20. Summer-like day; cool evening and night.
21. Cloudy weather with showers.
22. " "
23. " " Snow fall reported in the Bonnechere district; max. temp. 54 deg.
24. Bright morning; cloudy afternoon; cold evening and night; max. temp. 58 deg.
25. Cool, cloudy morning; cold rain towards 1 p.m.; showers during afternoon; cold evening and night.
26. Cool day, with showers; has been cool and cold now for one week; max. temp. 61 deg. 7m.
27. Warm day; fine, summer-like weather again; max. temp. 65 ° 7′.
28. Sultry hot day; max. temp. 76 deg. 3m.
29. Warm, oppressive day; smoke in air; another dry spell; max. temp. 78 deg. 2m.
30. Hot, hazy weather; smoke in air; gale of wind during night; max. temp. 79 deg.
31. Cloudy day; warm; max. temp. 72 deg. 7m.

JUNE.

The early part of June was very warm, and until the 10th the dry weather which had characterized the spring months continued. After the fall of rain on the 10th—the first of a succession of showers which lasted until the 19th—a relapse occurred in the weather, during which flurries of snow were reported, and severe frosts visited many sections of the country, causing considerable damage to crops. An extract from the Pontiac *Advance* says: "The two nights of frost have done more harm than was at first anticipated; the bean crop has been completely killed; potatoes, too, in some localities have suffered severely." Numerous other despatches reached us with regard to these frosts, one of which, from Charlesburg, reports a sharp frost and light fall of snow on the St. Louis road, realizing Mr. Vennor's probability where he says, "I shall not be surprised should there be an approach to snow in these midsummer months." It has certainly been acknowledged as such by most people. The Kingston *Whig* of the 25th says: "We are having renewed faith in Vennor. He made mistakes last spring, but no man is infallible. He is reading the present weather very correctly, however." Another feature of this month was a very disastrous storm on the 29th, the results of which were telegraphed from all parts of Western Ontario. Hailstones the size of walnuts covered the ground to the depth of several inches, to which window panes, roofs, tomato and potato vines fell victims, while the lightning struck down houses, barns and trees, and killed valuable cattle. Other severe storms were telegraphed during the month. A despatch from Halifax on the 27th says, "The house of William Guy was struck by lightning a few days ago, and Guy was instantly killed." On reference to the daily record it will be seen that the rains predicted for this month fell, and the general outlines of the weather were correct.

MONTREAL RECORD.

1. Hot, hazy morning; no signs of rain; therm. 83°
2. Intensely hot, dry weather; rain much needed; unusual amount of hot weather so far.
3. Hot day; windy; thunderstorm at midnight; very little rain.
4. Cool, fresh morning; warm day; heavy rains in Manitoba and elsewhere to westward; spittings of rain in the evening; night cool; therm. 67° 8′.
5. Bright, cool morning; cloudy afternoon and evening; ther. 55° 5′
6. Rain still keeps off; fine, cool day; therm. 70° 2′.
7. Sultry, clouded morning; rain towards 4 p.m., of brief duration; evening and night very oppressive.
8. Hot, cloudy morning, with appearance of rain.
9. Very sultry day; cloudy; shower of rain towards 2 and 4 p.m.; very oppressive evening and night; therm. 79° 7′.
10. Rain last night and this morning; thick, muggy atmosphere; showers becoming more frequent; rain again towards 10 a.m.; steady rain up to 2 p.m.; windy and clearing at 3 p.m.; clear, cool evening; heaviest rain of the month so far.
11. Bright, cool morning; northerly wind; cold last night; evening and night again cold; fires comfortable.
12. Bright, cool morning; rained heavily evening and night; thunder and wind.
13. Cool, cloudy day; clear evening.
14. Bright, cool day; cool evening.
15. Fine, cool weather; thunderstorm and rain at 2 p.m.; sultry evening, with great display of lightning and distant thunder.
16. Showery, overcast morning; rain everywhere; much cooler; cool evening.
17. Cool morning, with fresh wind; cool evening and night.
18. Fresh morning; warm day with haze; Ottawa papers contain reports of hail and wind storms damaging crops; maximum temp. 78° 8′.
19. Heavy rain during night; high wind in morning; evening cold; night still, clear and cold, with frosts; potatoes and other plants more or less injured.
20. Severe frosts last night; warm day; cool evening; heavy rain during night.
21. Cold morning; cloudy; rained heavily towards 11 a.m.; thunderstorm towards noon, with deluge of rain; cold wind evening and night; decided relapse in the weather.

22. Cold, windy morning, with cold showers; almost cold enough for snow; N.W. wind; cold showers all day; evening fine and clear; night very cold, with severe frosts.
23. Bright, clear, cold morning; very severe frosts last night, much injuring plants; dark, still, cloudy evening and night.
24. Wet morning; day altogether fine and warm; light shower during evening.
25. Hot, windy morning; storm of wind and rain during afternoon; calm, fine evening.
26. Warm day, with fleecy clouds; warm evening; bright moonlight night; therm. 67°.
27. Hot day; clear, cool moonlight night; therm. 70° 3'.
28. Sultry day, with showers during afternoon; cool, cloudy evening and night.
29. Morning sultry and threatening; heavy thunderstorm at noon, and continuing to 3 p.m.; showery afternoon and evening.
30. Hot morning; threatening rain; heavy rain towards 5 p.m.; papers contain notices of flurry of snow along St. Louis Road, Quebec, during period of recent frosts.

NOTE.—The highest readings of thermometer only are given.

JULY.

July, 1877, was the stormiest on record for a number of years. On the 5th a thunderstorm occurred at Buckingham, during which a daughter of Hugh McNully was killed, and four days later the spire of the Methodist church in Prescott was struck by lightning. At Galt, Ont., a disastrous storm occurred on the 9th, by which a number of houses were unroofed and trees torn up by the roots. A despatch of the 9th from Pensaukee, Wis., says "a terrible whirlwind struck the town on Sunday night, leaving but three houses standing; the Gardner House, a large hotel, was demolished; residences, saw-mills, trees, fences, etc., were swept away; six persons were killed and twenty wounded." In the *Witness* of the 10th it was published that "the captain of a fishing schooner, who arrived up from the Gulf on Saturday, reported that some days ago several fishing schooners off Gaspe were surprised by a sudden and heavy storm; several of the vessels were driven ashore and wrecked, and the crews drowned." An Ottawa telegram of the 16th reported that the City Hall bell had been struck by lightning, and the Toronto *Globe* of the same date contained particulars of a terrible storm which swept over Teeterville, Waterford and Port Jervis, N.Y. A furious storm from the west visited Quebec on the 17th, causing considerable damage in port, and on the following day intelligence arrived from Stroud of one of the heaviest hailstorms ever witnessed there. Church windows were smashed, fruit trees stripped, and the fall wheat was threshed out, the ground being covered with grain. We might quote from numerous other sources regarding the storms, but pass on to notice another feature of the month. An earthquake was felt at Murray Bay on the 17th, and extended to Kamouraska and Quebec, startling people from their beds, and on the 23rd tug steamers arriving up from below reported a tremendous commotion in the river about the time of the earthquake; vessels rocked to and fro with great violence; sailors, with frightened looks, rushed on deck and hailed each other to ascertain the cause of such an unaccountable occurrence. In conclusion, we may be permitted to quote from the Peterborough *Examiner*, which says: "Vennor's midsummer weather predictions are very nearly correct, and those who began to doubt him seem now to have faith in his prophetic visions. He is, without doubt, the best weather prophet in the business."

COUNTY OF OTTAWA RECORD.

1. Thunderstorm in early morning; heavy rains in the afternoon, and gale of wind from westward; rains increasing in frequency.
2. Morning showery; clear at noon; cool evening.
3. Bright, hot morning; calm; heavy showers to north-westward; windy; night dark and stormy.
4. Very hot day; 80° 6′ in Montreal; close sultry evening and night.
5. Hot morning; heavy rain, with thunder and lightning, at 1 p.m.; showers all afternoon; raining hard at 5 p.m., with lightning and thunder; cool evening; girl killed by lightning at Buckingham.
6. Fine, warm morning; fleecy clouds; afternoon hot; cool evening and night; northern lights at night.
7. Bright, warm morning; fleecy clouds; very sultry with increasing clouds.
8. Shower this morning; very hot day; severe thunderstorm towards 4 p.m., with tremendous rain-fall; fine, cool evening.
9. Intensely hot day; 80° 2′ in Montreal; storms gathering on all sides; very heavy rain, with thunder and lightning, at 4 and 7 p.m.
10. Slightly cooler; cloudy; westerly breeze; fine, cool evening.
11. Very warm again; steamy clouds; S.W. wind; spittings of rain; fine, cool evening.
12. Sultry, overcast day; S.W. wind; heavy shower at 2 p.m.; cool, almost cold, evening and night.
13. Clear, cold morning; N.W. wind; spitting clouds in the afternoon; cold evening and night, almost frost.
14. Fine, hot morning; clouds increasing towards evening.
15. Rained during night and early morning; sultry and clouded evening and night, with rain; heavy rains at several places in Ottawa valley.
16. Dull, oppressive morning; intense sultriness in the afternoon; thermometer at Quebec 110°; dark, cloudy night.
17. Same cloudy and oppressive weather; heavy rain towards noon—the heaviest of the season so far; clearing afternoon; fresh westerly wind; great rain-storm at Quebec.
18. Sultry, clouded day; distant thunder heard.

19. Rained hard all morning; east wind; clear noon; high wind from S.W. evening and night.
20. Clouded, windy day, with storms to south-eastward and north-westward; cool, windy night.
21. Cold morning; north-west wind; heavy cold rain towards 7 p.m.; very fall-like evening.
22. Bright, hot day; thunder clouds in the afternoon; cool, fine evening.
23. Cold last night; heavy dew; hot day, with strong N.W. blow; cool, almost cold, evening and night; thermometer 81° 2′ at Montreal.
24. Calm, cloudy morning; hot, clear afternoon, with scorching wind; hottest day of season so far; evening very oppressive; therm. at Montreal 84° 4′.
25. Very hot again; no clouds; west wind; evening cloudy, with southerly wind; warmest spell this summer; therm. 88° 5′ at Montreal.
26. Same terribly hot and oppressive weather; calm; evening very sultry, changing suddenly to cool night; therm. at Montreal 88° 5′.
27. Cool, cloudy morning; sultry day; cool, cloudy evening; poured rain all night.
28. Sultry, clouded morning; everything steaming with muggy heat; day and evening showery and very sultry.
29. Same hot, muggy weather; rained heavily during forenoon; distant thunder; a great deal of rain fell to-day; thermometer at Montreal 84° 8′.
30. Clear, bright, warm morning; cool evening and night; therm. at Montreal 82° 9′.
31. Clear, cool, almost cold, morning; warm day; clear, cold evening and night, with great display of stars and aurora.

NOTE.—This month's record corresponds strikingly with that for the same month of the year 1875.

AUGUST.

The August of 1876 was the driest, perhaps, in a score of years, in striking contrast with which the month in 1877 was perhaps the most rainy, as a few newspaper extracts will show. The Woodstock *Review* of the 9th says: "The prospects for an abundant harvest in Innerkip were excellent, but the heavy rains of last week have done great damage to the grain that was not gathered in. Vennor's probabilities have been to some extent realized;" and on the 10th we hear from another source that "a hailstorm of unusual severity passed over the lower part of Allumette Island. It took a breadth of from one half to one mile, and levelled everything in its course, cut down the growing grain, corn and potatoes, and stripped the forest trees bare of their leaves." About this date hailstorms extended from Ottawa city northward to the Desert settlement, 90 miles up the Gatineau, as well as a long way beyond Pembroke. Mr. Vennor can attest to the fury of these hailstorms, having had an unpleasant experience in one. While driving forty miles north of Ottawa a terrific storm occurred, the hailstones, which were as large as plums, pelting down upon himself and his horse until the animal, smarting from their effects, refused to move, and the driver could only shield himself as well as possible, and anticipate fine weather after the storm. The hailstones, during the storms which were of daily occurrence between the 10th and 15th, frequently lay on the ground for twenty-four hours without melting, and the crops suffered severely in many sections of country. An Ottawa telegram on the 13th read that "A farmer from the mountain district says that some of the hailstones were as large as good sized hens' eggs, and that one of them struck an employee of his in the face, and inflicted a severe wound. Fowls were killed, and travellers were obliged to take refuge in barns and other places." In North Onslow the storm had no parallel in the memory of the "oldest inhabitant," and many thought "the end" was at hand. On the 15th a very severe storm of thunder and lightning was telegraphed from Ottawa, and a newspaper correspondent at Kazabazua, on the Gatineau river, says that district was visited by a "most violent and destructive storm. At Aylwin, about four miles above Kazabazua, the school house was struck during the dinner hour, whilst the children were playing around in little groups. The accident caused the wildest confusion among the children, some of whom fainted away. The windows were all knocked out, the blackboard torn off the wall and set on fire, the beams were shattered, and one of the boys, named Orr, was violently thrown to the floor, remaining insensible for some time. A building owned by Hamilton Bros., in the same vicinity, and occupied by Eli Emery, was damaged. Mrs. Emery, who was washing dishes at the time, and her daughter received a severe shock; the heels of the girl's boots were torn off and her legs badly scorched. A son of Mr. George Hartley was knocked insensible. The Guelph *Mercury* of the 17th says: "Many farmers will lose heavily by the past heavy rains—a very unfortunate affair after having repeated light crops during the past few years;" and an Ottawa despatch of the 18th says, "A farmer estimates that at least eight bushels of peas to the acre were threshed out by the hail during the recent storm." Very heavy storms occurred at intervals during the whole of this month, and one particularly worthy of note on the night of the 29th passed over Montreal, the electric fluid entering the Central Police Station, and shocking Sergeant Neilson and a son of the caretaker in a manner which they have cause to remember.

RECORD FOR THE GATINEAU AND LIEVRES SECTION.

1. Warm, sultry day, with southerly wind; ther., Montreal, 85° 3'.
2. " " " 80° 5'.
3. Warm day; cool evening and night; ther. at Montreal 80° 6'.
4. Cool day; strong northerly blow in afternoon; clouded, cool evening; very fall-like weather.
5. Cool morning; north-west blow; wind and rain in the afternoon; unsettled, fall-like day; thunder-storms coming from N.W.

6. Beautiful, cool and slightly clouded day ; rain during night.
7. Rain last night ; unsettled day ; cold evening, with N. W. wind.
8. Cool, clouded day ; thunderstorms to northward ; great display of lightning during evening and night.
9. Wind from eastward ; heavily clouded ; severe storm during afternoon, with a great deal of thunder and lightning.
10. Sultry morning, with easterly wind ; clouded, hot day ; thunder-storms tc northward very severe.
11. Thunder-storms and intense sultriness all day ; wet evening ; storms very severe to-day and on the 9th all up the Gatineau ; great hail-storms in some sections, resulting in considerable damage to grain.
12. Sultry weather, and thunder-storms in all directions ; heavy settled rain during afternoon and evening.
13. Clouded and sultry ; storms in all directions ; cool evening.
14. Sultry, clouded morning ; intensely hot afternoon and evening, with a great deal of lightning ; severe thunderstorm towards midnight, lasting for hours, and extending over a large area of country ; heavy rain ; great hail-storm in March and Eardley, on the Ottawa river ; great destruction of crops.
15. Hot and clouded day ; evening clearer and cooler.
16. Rained more or less all day ; cold, heavy rain during afternoon ; dark, unsettled evening ; very wet night.
17. Poured rain all last night and this morning ; cleared toward noon ; cool, almost cold, night.
18. Bright, pleasant weather ; first fine settled weather for harvest.
19. Clouded morning ; heavy shower in afternoon ; evening fine.
20. Another fine day for harvesting ; dry heat.
21. " " "
22. Very sultry and oppressive day ; evening clouded and stormy-looking ; therm. at Montreal 86° 1'.
23. Terrible heat ; hottest day of season so far ; thermometer readings very high everywhere ; thunder and lightning, with some rain, during the evening ; max. temp. at Montreal 88°.
24. Overcast and cooler ; southerly wind ; light rains ; rained all night in Ottawa and vicinity.
25. Warm and clouded day ; thunder-storms seen passing to north-eastward ; evening decidedly cool.
26. Fine, warm and slightly cloudy day ; heavy thunderstorm to northward during evening ; evening and night cool.
27. Cool, breezy morning ; clouded afternoon ; evening and night dark, with light rains.
28. Sultry day, with steamy clouds ; clouded evening and very sultry ; rain set in at 10 o'clock p.m., and continued most of the night.
29. Bright, warm morning ; steamy clouds ; thunder-storm with rain at 1 o'clock p.m. Storms during afternoon and evening.
30. Bright, beautiful day ; fleecy clouds ; cool evening and night.
31. Cloudy day, with easterly wind during forenoon ; dark afternoon ; wind from south-westward ; cold evening, with heavy blow from north-westward during night ; heavy rain at Toronto.

SEPTEMBER.

MONTREAL RECORD.

1. Morning cloudy and cool ; afternoon bright and pleasant ; wind N.E.
2. Morning cool ; indications of rain in the afternoon ; thermometer at 7 a.m. 53°.
3. Cold and cloudy.
4. No change in the weather ; light rain-fall ; water gradually falling.
5. Rain fell from 8 a.m. to 5 p.m. ; wind variable.
6. Clear, cool and pleasant.
7. Bright, pleasant day.
8. No change.
9. Fine, seasonable weather ; wind variable.
10. Early morning dull ; bright and pleasant at noon.
11. Fair and pleasant.
12. Cloudy ; atmosphere close and warm.
13. Fair, clear and warm ; wind from S.W. to N.W.
14. Close and warm ; morning foggy ; clear at 7 a.m.
15. Still warm ; fog in the morning.
16. Warm ; light shower of rain in the afternoon.
17. Bright and clear ; very warm ; heavy rain from 9 to 10 p.m., with high wind from N.W.
18. Fair and pleasant.
19. Morning clear ; afternoon cloudy.
20. Cloudy and cool ; rain at 10 p.m.
21. Morning cloudy ; afternoon clear and pleasant.
22. Bright and pleasant ; water falling.
23. Fair weather ; fresh breeze from the west in the evening.
24. Weather seasonable.
25. Morning fair ; noon cloudy ; heavy storm at 6 p.m. ; vivid lightning and heavy rains.
26. Morning clear ; afternoon cloudy ; west wind.
27. Clear and pleasant.
28. Early morning clear ; heavy rain at 11 a.m. ; afternoon showery and sultry.
29. Morning foggy ; cloudy day.
30. Fine, bright day ; wind westerly.

OCTOBER.

MONTREAL RECORD.

1. Cloudy and warm ; thunderstorm at 10 p.m.
2. Cool and cloudy ; water rose an inch.
3. Cold and cloudy ; rain between 8 and 10 p.m.
4. Dull and cloudy ; wind variable ; rain from 10 a.m.
5. Cool and pleasant.
6. Fine, seasonable weather.

7. Clear and bright; wind W.N.W.
8. Cloudy and dull; wind S.E. by S.; rain at 6 p.m.
9. Raining all day.
10. Foggy morning; cloudy day; evening fair.
11. Rain all forenoon; afternoon unsettled.
12. Rainy all day; wind N.E.
13. Showery; wind easterly.
14 Cloudy and showery; very unsettled weather.
15. Dull and cool; showers during the day.
16. Heavy rain last night; showery morning; afternoon clear.
17. Bright and clear.
18. Afternoon cloudy, with variable winds.
19. Cloudy and dull; light rains at times.
20. First frost last night; cloudy and dull.
21. Cool and cloudy.
22. Variable wind; weather unchanged.
23. Light rains in the morning; cloudy day.
24. Morning foggy; rained all afternoon; wind N.E.
25. First snow of the season; wintry appearance; cold and disagreeable.
26. Clear and cool.
27. Bright day, with variable winds.
28. Morning foggy; clear at 9 a.m.
29. Rain in the morning; cloudy at noon.
30. Weather clear and bright after the rain.
31. Morning hazy; rain at 10 a.m.; wind south.

TORTOISES AS WEATHER INDICATORS.—If there be any truth in a paper read by a French *savant* at a recent meeting of the Academy of Sciences in Paris, every well-regulated household should have one or more tortoises about the premises. According to M. Bouchard, tortoises take extraordinary precautions against the cold weather. Their instinct tells them in the milder seasons when the thermometer is likely to fall to freezing point, and toward the end of autumn warns them also of the approach of winter. In both cases they take precautions to screen themselves from cold, and by carefully observing them M. Bouchard has for years been enabled to regulate his hot-house. At the end of autumn, when the winter threatens to be severe, tortoises creep deep into the earth, so as to completely conceal themselves from view. If, on the contrary, the winter promises to be mild, they scarcely go down an inch or two—just enough to protect the openings of their shells. Last January, which was so mild, they even went about. Last month, the thermometer standing at 50° Fahr., our author saw his tortoises creep into the ground, and that very night the glass fell to 28° Fahr. On the 1st inst., the mercury being at 110° Fahr. in the sun, one of the tortoises hid itself; on the following morning there was hoar frost.—*Forest and Stream.*

THE EARTHQUAKE OF NOVEMBER, 1877.

The earthquake of November 4th, 1877, was felt all through the eastern section of New York, New Hampshire, Vermont, western Massachusetts, and the province of Quebec north and west as far as Ottawa, and east to Sorel. In Montreal it was felt at precisely between ten and eleven minutes to two a.m., the motion being from eastward to westward. The shock lasted about thirty seconds, the premonitory rumbling occupied perhaps another twenty seconds, and the receding noise some thirty seconds more. The Montreal *Witness* describes it as follows :

"The first announcement of the disturbance was a low rumbling sound, which perceptibly grew harsher, ending with what might be termed a bumping or explosive noise ; then came the shock or tremor, which was felt most severely by those sleeping in the upper stories of tall houses.

"Some report two shocks, one following the other, and the *Minerve* says there was a second about half an hour after. In one case, in a bank, quite a swaying motion was experienced by the messenger, who immediately rushed down the lighted staircase, imagining that burglars had lifted the safes and vaults with a ton of dynamite or so. When he had got into the dimly-lit basement all was silent and still as the grave, and though he says cold chills didn't run over him, yet he admits that he felt frightened. The majority of heavy sleepers, judging by several cases which have been noted by our reporters, were not wakened by the shock ; but there must have been thousands, however, who were, because the general inquiry in the morning, 'Did you feel the earthquake last night?' was nearly always answered in the affirmative.

"No casualties are reported, except that a great many people got very much frightened, imagining that burglars or supernatural visitors had intruded themselves. The *Herald* says a gentleman residing on St. Catherine street west, who was reading at the time, was thrown three feet from his chair, and numerous similar instances are reported, the distance in each case being greater than in the previous one ; and the *Gazette*, not to be outdone, announces that the Queen's statue in Victoria square swayed her sceptre for once, and seemed to the astonished policeman on duty there to be beckoning the Bonaventure block to fall upon him and wipe him out of existence.

"These remarkable manifestations of disturbance seem to stand out isolated and alone, and as far as we can learn nothing but the tongues of bells were moved and doors slightly swung. Of course, to those out upon the streets at that hour, in possession of all their faculties, the phenomenon was invested with a painful air of reality. A policeman at the west end said he was walking slowly along the stone pavement, when suddenly he thought that he heard a street car being driven rapidly towards him, and was actually wondering why he could not hear the tinkling of the bells on the horses' necks, when he was greatly unnerved by feeling the ground undulating under him. He ran out from the walk into the middle of the street, 'For,' said he, 'the block of three-story houses back of me was all of a shake, and the pavement was fairly rolling toward me, like as if you would shake a carpet.' A gentleman on St. Catherine street, who was roused, went to the window and perceived a policeman standing in the middle of the street, and apparently in a state of terror. In response to a question, the alarmed guardian of the peace remarked that 'something was up,' but he could not exactly say what."

"The reports show that the shock passed underneath all parts of the city, and was felt by the sailors on watch on board the ships in harbor. Upon the direction of the shock being with the parallels of latitude all agree, but as to whether it went from west to east or east to west is disputed. It is said that dogs were observed to be uneasy, and endeavored to get outside of buildings. The night was windy at first, but more serene and star-lit when Nature began to travail. The temperature was not unusually low.

"A gentleman living on Drolet street gives the following account: 'I was awake at five minutes before two o'clock at the time of the earthquake, and noticed the first indication at about that time. It was as if a heavily loaded wagon was passing along. This noise then ceased, and in a few seconds it recurred, increasing to a shock like a rupture, then decreasing again, the time occupied being one and a half minutes. I may say that, my door being ajar, I could count the vibrations, which were east and west.'

"A gentleman residing on Drummond street says that the duration of the disturbance after the shock could not have been less than forty-five seconds, as during its progress he was enabled to rise leisurely and strike a light. It was nine minutes to two by his watch (subject to correction of time) when the noise was done.

"A gentleman in the west end says the shock came with such violence as to make him fear that there was some danger of the house falling down.

"Two different persons stated that they heard a rushing or a fizzing noise accompanying the shock. In one case this might have been caused by telegraph wires clashing together, but it could not have been from this cause in the second case.

"A conjunction of the planets Mars and Saturn occurred about the hour of the earthquake, and whether this may have had any connection with the quake is for scientists to determine. It is well known, however, that Mercury is connected with fully one half of the worst of our meteoric disturbances. Mansill says: 'The principal disturbing positions of the planets for November appear to be located about the 3rd, the 10th to 13th, and the 20th to 21st.' There was also a new moon on the 4th."

MY OWN CHAPTER.

THE AUTUMN, WINTER AND SPRING OF 1877-78.

Our birch canoes were quietly gliding down the last stretch of the Rivière aux Lievres, and the sullen boom of the falls at Buckingham had just broken in upon the quiet of a sultry evening in September, when we were hailed by a solitary canoeman from the middle of the stream, "Letters for Vennor and party!" Vennor and party were instantly on hand, and shortly each one had received his batch of "home correspondence." In my own packet my eye was at once arrested by a very familiar monogram, namely, that of my indefatigable weather clerk, and eager to learn the latest weather sensation, I opened and perused the letter as we paddled over the short distance which still intervened to our landing place. And this is what I read: "Well, we are waiting for your weather predictions for the autumn, winter and spring of 1877-78. What are they to be? Printers are bothering my life out, and I wish to have a little left to finish up your Almanac. Hurry up! Yours truly, etc." It came upon me as a thunderbolt. Was there to be another almanac? Had I really promised another forecast of the weather for a year in advance? Instantly visions of Toronto *Globes*, Montreal *Gazettes*, lurid *Stars* and Bobcaygeon *Independents* floated through my brain, while in the mirror-like waters of the Lievres I fancied I saw pictured a grinning *Grip* and hideous cartoon. I had taken the advice of the *Witness* poet in

THE SEER'S LAMENT.

I know the covert of the wolf,
  The red deer I out-ran,
The heron's haunt and the snowy lair
  Of the wintry ptarmigan.
The owl was my familiar,
  I knew the beaver's plan,
And the garter snake he whispered me
  What is not known to man.

From my deep Laurentian Valley
  I pierced the infinite blue,
Bathed in the dewy influence
  Of Pleiades I grew.

I knew Arcturus and his sons,
Orion bold I knew,
And the epochs and the omens
Of Sirius' changing hue.

I loitered with the zephyr
On the balmy summer morn,
And scampered with the hurricane,
And knew where he was born.
The equinoctial whirlwind
Full oft I laughed to scorn,
For I knew his inmost secrets,
And his terrors I had shorn.

I pierced the foggy treasure house,
And the wealth of waters told;
I scanned the stores of Winter,
And measured out the cold.
I have walked with old December
And his burly brethren bold,
And I made a winter almanac,
And twenty thousand sold.

I drove the chariot of the wind,
I gave the clouds their form,
And oft arrayed at my behest
The armies of the storm;
And all their grand artillery
Would soldierly perform;
Should I not tell the people
The cold days and the warm?

I told the snow its seasons,
And the frost its setting in;
In letters to the *Witness*
Bright laurels did I win.
And multitudes believéd,
And counted it a sin
Not to believe the thaws and dips
That I had writ therein.

But days of evil followed,
And the shadow back did roll,
And the unchained powers of nature
Took vengeance on my soul,
When my prophetic Python turned
To me his negative pole,
Nor could I bid him silence,
Or his lying tongue control.

For March the very ancients
Had figured by a ram—

My demon brought him on the boards
  A lion and a lamb.
And the people chaffed my lion,
  And called the lamb a sham,
And my St. Patrick's prophecy
  An unexampled "cram."

Then all the minor prophets crowed
  With wings and crest elate,
And all the dogs of jealousy
  Barked out their little hate.
Oh, would that I could but be still
  Till this is out of date,
And next year's winter almanac
  Might reconstruct my fate.

But now, notwithstanding my faithful adherence to this advice, as surely as the recurring seasons my time of trial had again arrived. What was to be done? In my canoe were three weather-beaten *voyageurs*—men who from infancy had "paddled their own canoes." Turning to these, I abruptly demanded:

"What kind of an autumn are we going to have?"

"Plenty rain," "Not much water," and "Dunno," were the replies I received.

"What sort of winter shall we probably have?"

"Oh, plenty good deep snow—some good cold," "Pshaw! Great plenty rain—dry summer wet winter—not much cold," and "Dunno," were the respective answers.

Once more I ventured, "Then we shall have a dry spring?"

"No, wet—plenty wet," said the one.

"Pshaw! Wait, you see. I bet you not much water—plenty rain byme-by next summer," cried the second.

"Yer two durned fools!" growled the third, "and dunno nothing. Ain't you seen the beavers?"

"No! What you see?" eagerly questioned my two Frenchmen.

"You ain't seen the beavers?" again reiterated No. 3.

"No, no! See no beaver. *Pourquoi*, eh?" gasped the expectant Frenchmen.

"Just case I hain't seen a durned one myself for more'n a twelvemonth," quietly grinned the pilot, as, with a vigorous spurt of the paddle, he brought the canoe up to the landing place.

Comforting myself with the assurance that perhaps, after all, I knew more about the weather than the whole three put together, I stepped ashore, but not until I had quietly whispered in the ear of the "rainy" Frenchman,

"Plenty rain, eh? Warm autumn? Wet spring, eh? Come along—you're my man. Great minds think alike."

Thus did the shadowy skeleton of my almanac for 1877-78 first present itself to me; but I hurry to add that this skeleton ere long assumed a more definite shape, became clothed with flesh, muscle and skin, and finally stood forth so life-like and natural that I again entrusted it with my reputation as a weather-cock, and have actually persuaded myself that in spite of "Pshaw!" the people of Canada do really believe that Vennor, after all, does know something about the weather.

THE WEATHER TO COME.

Shortly after the canoe episode just related, and about the 1st of October, I despatched to my Almanac publishers my first hurried impressions of the weather for the autumn of 1877 and winter and spring of 1878. These read as follows:

"Indians, farmers, trappers and lumbermen whom we have interviewed over a very broad-spread area all agree on one point, viz., *great precipitation* during the fall and winter months approaching, but whether this will be as snow or rain is a question I found few prepared to answer definitely. This great precipitation is but a natural conclusion to arrive at, for otherwise we should of necessity have a most unprecedented condition of things. The majority of the people interviewed are inclined to believe that deep snows will be the programme for next winter, while the minority prognosticate rains and very open weather. With this last party I myself fall in, and for some such reasons as the following. When there is great precipitation (snow or rain) during any particular season the temperature is seldom very low. Therefore, should this first and leading impression be correct, we may expect a moderate winter. Further, as thunder-storms have been unusually prevalent throughout the whole of the past summer, and even still continue, (1st Oct.), I look for a temperature more productive of rains than snows. Consequently our approaching winter and spring will, in all probability, be moist and slushy. But I expect some sharp seasons, and and the first of these will come early. Snow will probably fall early, but will not remain.

"*October* will be a cold month, with snows and rains.

"*November* bids fair to be warm, with but few severe frosts, until towards its latter part, and I am inclined to locate in this month a well marked and beautiful Indian summer.

"*December* will in all probability set in very sharp, but this cold term will be of short duration, and give place speedily to rains and snows.

"*January*, of 1878, looks to me at present gloomy, wet and foggy, and not unlike that of the season of 1875.

"*February* again—I must be cautious about this fickle month this time—will set in severely. I look for more snow than rain; consequently this month will be probably the most wintry-like of the whole winter.

"*March* will bring more snow during its early part, but the month will end wet, with heavy winds, and bids fair to go out exceedingly stormy.

"*April*, *May* and the first half of *June* will be very wet. This impression is firmly imprinted on my mind, and this wet spring will probably be followed by an intensely hot, muggy midsummer.

"The whole autumn and winter will be favorable to the increase of throat diseases and fevers, also cattle diseases, and I agree with Prof. Mansill in anticipating the approach of Asiatic cholera towards northern latitudes.

"*Rivière aux Lievres*, Oct. 6th, 1877."

Such was the sketch hastily penned and sent off to my printers early in October, and I have only to add that three subsequent forecasts made from other and more extended data agreed so closely with the first in general details that I have determined to abide by this, and merely intend to paint in slightly firmer colors a few features in the weather which may be particularly noticeable.

Indian Summer.—I am inclined to predict a warm and well-marked Indian summer in the month of November, as I believe October will be cold and wet.

Cold Snaps.—The first of these I would locate in October, but of course this will be moderate. The second will arrive towards the latter part of November or early portion of December, and waters will become pretty well ice-locked. This also will be of short duration, giving place to heavy rains and snow-falls with open weather. A third cold term will probably enter with or close upon the entry of February, and this I am inclined to sketch as more protracted than the preceding ones. It will moderate to heavy snow-falls rather than rains.

Thunder-storms.—Judging from the action of the thunder-storms this year, I should expect these to continue up to a late date, and it will not surprise me should I have to record one or more during the approaching winter.

Earthquakes.—There is now going on in the long range of Laurentian Mountains a very considerable agitation. On several occasions while camped out this summer we have felt distinct shocks and tremors of the earth. "Trembling Mountain," in the "Nation" waters, and "Devil Mountain," between the Gatineau and Lievres rivers, have scared the Indians for miles around by their hollow rumblings; while, more recently, an earthquake wave has swept over a large part of North America.

H. G. V.

CHOLERA—PROF. MANSILL'S PREDICTION.

Prof. Mansill's "Almanac of Planetary Meteorology" for 1876 contains the following remarkable prediction in regard to cholera, which is being partially fulfilled:

"We may expect the next regular cholera epidemic period to commence about 1876 or 1877 in southern latitudes, and reach its height about 1878 or 1879 in northern latitudes, and return to southern latitudes again about 1880. * * There will probably be cholera epidemic in the East Indies, and perhaps further west, during the spring and early summer months of 1876."

Showers of Toads.—Notwithstanding the fact that the so-called showers of toads have been accounted for by naturalists showing that the long needed rain has called myriads of young toads from their hiding places, and the ground where none were seen a few hours before suddenly becomes alive with the little creatures who come forth to enjoy the moisture, there are still many who firmly believe in their fluvial origin. There is a fact in this connection that does not appear to be generally known, even to those who are well posted in such matters, which is, that the young toad has two modes of development. The best known one is that of passing through the tadpole state when the eggs are laid in water, in a manner similar to that of the frog; the other is the wonderful property that is possessed by the egg of a toad, enabling it to skip the tadpole form and hatch a perfect toad if laid in moist earth instead of water. It is not improbable that a warm rain may be required to develop the embryo, or at least release it from the egg. If so, a "shower of toads" is the result.

METEOROLOGICAL INSTRUMENTS.

THE BAROMETER.

Galileo, towards the close of his life, being asked to explain why it was that, if Nature abhors a vacuum, water could not be raised by a suction pump higher than about thirty-two feet, was compelled to admit that Nature's abhorrence was measured by a column of water that height. His thoughts being once directed to this question, he followed it closely until death put an end to his labors, but not before he had strongly recommended his pupil, Toricelli, to continue the investigation.

Toricelli, following up the matter, argued that the power which held up a column of water to the height of thirty-two feet would hold up a column of mercury—mercury being fourteen times heavier than water—a proportionate height. To test this, in 1643 he filled with mercury a glass tube about three feet in length, closed the open end with his finger, and inverting the tube, plunged it into a basin of mercury. On removing his finger the mercury sank till it stood at twenty-eight inches in the tube, leaving a vacuum at the upper end.

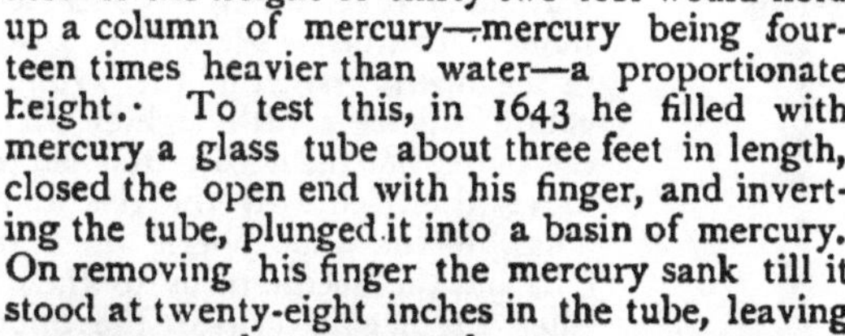

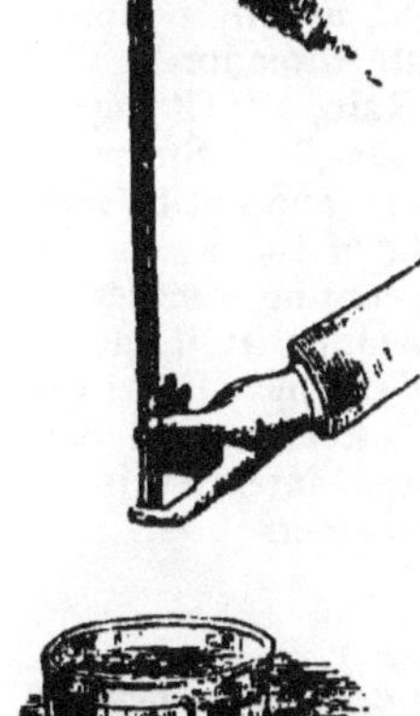

Continuing his experiments, he discovered that the mercury was sensitive, and rose and fell according to the condition of the atmosphere, the height of the former changing with the slighest variation in the weight of the latter. He, also, died before his observations were completed, but Pascal, in Rouen, France, took them up, and found that as the thermometer was elevated or depressed the mercury fell or rose in the tube, and that at the same moment the mercury in two barometers at different elevations would stand at different points. From this he devised the principle by which heights were measured by this instrument.

Otto Guericke, an ingenious and wealthy Madgeburg burgomaster, early in the history of the barometer constructed a gigantic one for indicating the state of the weather. It was a glass tube thirty feet high, nearly filled with water, which was erected inside the walls of his house, and rose above the roof. In the upper part of the tube, which was larger than the rest, was placed the figure of a man, so large as to be visible from the street. In fine weather this man, floating on the water, was raised above the roof, but on the approach of foul weather descended into the house.

The most perfect barometer of the present time was constructed by Prof. Daniell on the same principle as this one, and now stands in the hall of the Royal Society at Somerset House. The glass tube is forty feet long and an inch in diameter. The water in the tube stands on an average four hundred inches above that in the cistern. The column is sensitive to continual changes in the atmosphere which do not affect other barometers, and in windy weather vibrates up and down almost with the regularity of respiration.

Perhaps the most common form of barometer in Canada is the wheel barometer, in which the varying height of the mercury is indicated by the movement of a needle on a divided circular dial. This is accomplished by adopting the syphon form of the barometer tube, which is concealed behind the dial and frame. An iron or glass float, sustained by the mercury in the open branch of the syphon, is suspended by a centre balance a little lighter than itself. The axis of the pulley has the needle attached to it, and consequently moves the needle by the rise and fall of the mercury. Thus, if the atmospheric pressure increases, the float falls and the needle turns to the right, and if it diminishes, the needle turns in the opposite direction. The wording of these barometers of "Rain," "Change," "Fair," "Set Fair," "Very Dry," "Stormy," "Much Rain," is, of course, arbitrary and apt to mislead, as it is not as much the weight of the air as the *changes* in its weight which indicate coming weather.

In "taking a reading," it is important that it should be done as quickly as possible, as the heat from the body and the hand is sufficient to interfere with that accuracy which is necessary when the intention is to compare the readings with other barometers.

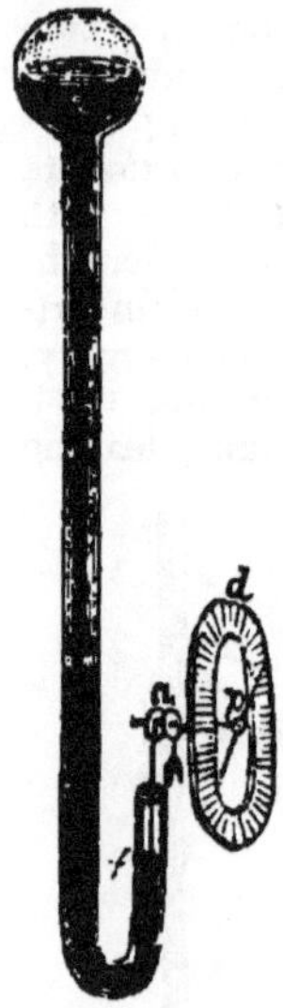

THE THERMOMETER.

The Thermometer is an instrument for measuring degrees of heat by the contraction or expansion of fluids in enclosed tubes. The tubes, which are of glass, have spherical, elongated or spiral bulbs blown on to one end; they have also an exceedingly fine bore, and, when mercury or spirit is enclosed in them, these fluids, in contracting or expanding with variations of temperature, indicate degrees of heat in relation to two fixed points, viz., the freezing and boiling points of water. In filling the tube, mercury, colored so as to be easily visible, is commonly used, the air first being excluded from the tube so that there will be a perfect vacuum, and thus no resistance be offered to the expansion of the fluid.

When the fluid (either mercury or spirit) has been enclosed in the hermetically sealed bulbous tube, it becomes necessary in order that its indications of elevation or depression of temperature may be comparable with those of other instruments, that a scale having at least two fixed points should be attached to it. Consequently, as it has been observed that the temperature of melting ice or freezing water is always constant, the height at which the fluid *rests* in a mixture of ice and water has been chosen as one point from which to graduate the scale. It has also been further observed that with the barometer at 29.922 the boiling point of water is also constant; and when a thermometer is immersed in pure distilled water, heated to the boiling point, the point at which the mercury remains immovable is, like the freezing point, carefully marked, and with these stationary points indicated, the tube is divided into as many equal parts as are necessary to constitute either of the three scales at present in use. These three are the Reaumur, introduced in 1730, the Fahrenheit, in 1749, and the Celsius, in 1742. The first is commonly used in Russia and the north of Germany, the Fahrenheit in England, her colonies, and the United States; and the last, commonly called the Centigrade, in France and the portion of Europe not previously mentioned. In the Fahrenheit scale the freezing point is 32°, and the boiling point 212°, so that the intervening space is divided into 212—32, or 180 equal parts or degrees. In the others the freezing point is the zero, but in the Reaumur scale the boiling point is 80°, and in the Centigrade 100°.

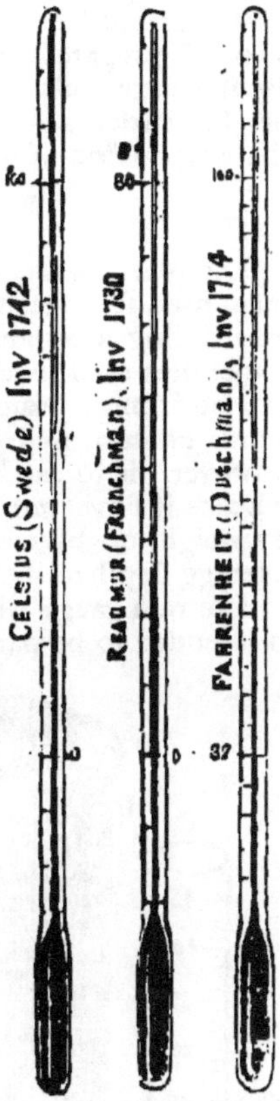

Celsius (Swede) Inv 1742
Reaumur (Frenchman) Inv 1730
Fahrenheit (Dutchman) Inv 1714

As a variety of circumstances arise in which it becomes necessary to convert readings from one scale into those of the other, the following rules are given:

1. To convert Centigrade degrees into degrees of Fahrenheit, multiply by 9, divide the product by 5 and add 32.

2. To convert Fahrenheit degrees into degrees of Centigrade, subtract 32, multiply by 5, and divide by 9.

3. To convert Reaumur degrees into degrees of Fahrenheit, multiply by 9, divide by 4, and add 32.

4. To convert Reaumur degrees into degrees of Centigrade, multiply by 5 and divide by 4.

THE HYGROMETER.

The amount of moisture in the air is measured by the hygrometer. The consideration that a certain amount of moisture in the air is necessary to the continuance of health suggests the advantage of maintaining a due proportion in the atmosphere of sick rooms, where the artificial heat, so often injudiciously used, disturbs the healthful hygrometic condition of the air. By this instrument the amount of aqueous vapour held in the air is effectually indicated, and by it many hints, which, if acted on, would prove of great value to the patient, may be obtained.

Hygrometers are based on three principles, indicating the presence of moisture by absorption, condensation, or evaporation. By the first class of instruments the indications are the result of the contraction or expansion of prepared human hair, oatbeard, catgut, seaweed, grass, &c. By the second class, the moisture is condensed on bright polished silver or glass surfaces. By the third class, the moisture is indicated through the evaporation of fluid in a bulb in proportion to the degrees of air.

THE RAIN GAUGE.

The pluriometer, or rain gauge, as its name indicates, is an instrument used for measuring the amount of rain which falls upon a given area during a certain space of time. An easily constructed rain gauge sometimes used, is a tub, or bucket with a thin edge, which is placed in a horizontal position for catching the rain, whose depth may afterwards be measured by a graduated rod. The more common method, however, is to catch the rain fall in an accurately made funnel, from whence it flows into a receiver of any shape. It is then either measured by weight or by means of a tall graduated cylinder, which gives the average depth of the rain-fall.

The rain gauge whose picture is given is one of the latter class. It is intended to be partly sunk in the soil to keep the contents perfectly cool, and the receiving surface of the funnel, accurately turned to a diameter of eight inches, terminates at its lower extremity in a curved tube, which, by always retaining the last few drops of rain, prevents evaporation. The graduated vessel in which the depth of the fall is measured in this instance is divided to 100ths of an inch, having due regard to the larger area of the funnel.

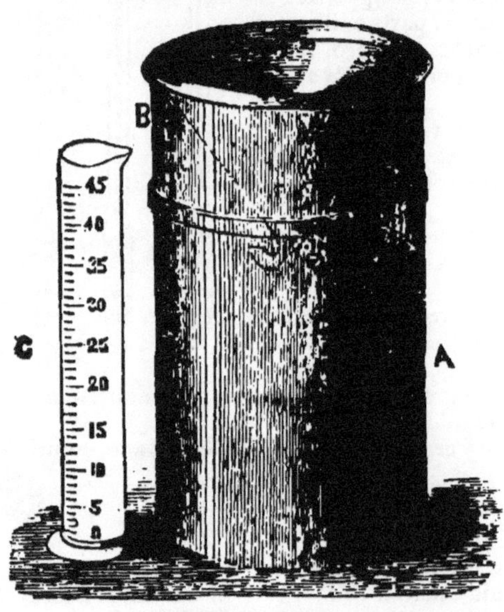

RAIN GAUGE.

It is difficult to employ a rain gauge to measure a snow fall, unless the air is perfectly still, as the wind interferes with the snow's reception. An easy and fairly satisfactory method of obtaining the snow fall is, after a snow storm, to take a cylindrical vessel of sufficient depth, and with it cut out a section of the snow from some place where it has fallen evenly.

The snow thus obtained may be melted, or dissolved in a known quantity of water, and the depth of the fall thus obtained.

Much care must be exercised in the placing of rain guages. They should not be placed in the neighborhood of trees and buildings, nor on the tops of isolated buildings. The standard position of the mouth of the gauge is from eight to sixteen inches above a broad level lawn.

The value of the rain gauge is well indicated by Luke Howard, in his "Climate of London." He says :—"It must be a subject of great satisfaction and confidence to the husbandman to know, at the beginning of a summer, by the certain evidence of meteorological results on record, that the season, in the ordinary course of things, may be expected to be a dry and warm one, or to find, in a certain period of it, that the average quantity of rain to be expected for the month has fallen. On the other hand, when there is reason, from the same source of information, to expect much rain, the man who has courage to begin his operations under an unfavorable sky, but with good ground to conclude, from the state of his instruments and his collateral knowledge, that a fair interval is approaching, may often be profiting by his observations, while his cautious neighbor, who waited for the weather to settle, may find that he has let the opportunity go by. This superiority, however, is attainable by a very moderate share of application to the subject, and by keeping a plain diary of the barometer and rain guage, with the hygrometer and vane, under his daily notice."

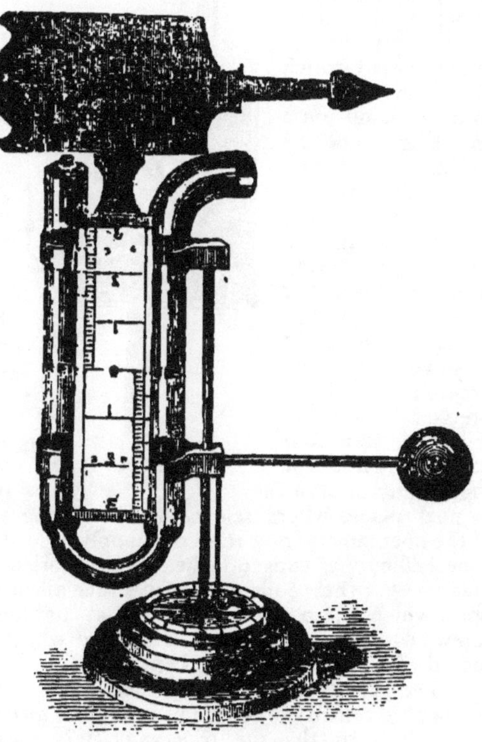

LIND'S ANEMOMETER.

THE WIND GAUGE.

The anemometer, or wind gauge, is an instrument used for measuring the force of the wind. One of the earliest forms consisted of a glass syphon, whose limbs are parallel to each other, and mounted on a vertical rod, on which it freely oscillates by the action of the vane which surmounts it. The upper end of one limb of the syphon is bent outward at right angles to the main direction, and the action of the vane keeps this open end of the tube towards the quarter from whence the wind blows. Between the limbs of the syphon is a scale graduated from 0 to 3 in inches and 10 lbs., the zero being the centre of the scale. When the instrument is used, it is only necessary to fill the tube with water to the zero of the scale and then expose it to the wind. The force of the wind on the water depresses it in the one limb and raises it in the other, and the depression and consequent elevation is the height of the column of water the wind is capable of sustaining at the time of observation. The contraction in the syphon at the lower bend is to prevent the jumping effect on the water in the tube which would otherwise be caused by sudden gusts of wind. Lind's anemometer, an engraving of which is given, is one of this class, and the following table shows the force of wind on a square foot for the different heights of the column of water as shown by it. Six inches represents a pressure of 31.75 lbs. to the square inch, or a hurricane; 5 inches, 26.04 lbs. pressure, a violent storm; 4 inches, 20.83 lbs. pressure, a great storm; 3 inches, 15.62 lbs. pressure, a storm; 2 inches, 10.42 lbs. pressure, a strong wind; 1 inch, 5.21 lbs. pressure, a high wind; .5 inch, 2.60 lbs. pressure, a brisk wind; .1 inch, .52 lb. pressure, a fresh breeze; .05 inch, .26 lb. pressure, a gentle breeze; 0, a calm. Many of our readers will have seen four hemispherical cups, such as are shown in the second engraving, revolving rapidly or slowly, as the case may be, in or near places where science has its resting place. They are portions of the anemometer now most commonly used. It consists of four hemispherical copper cups attached to two horizontal metal arms in such a manner that their concave surfaces face all one way. The vertical axis upon which these are secured has at its lower extremity an endless screw, whose motion causes a toothed wheel to revolve, by which the record of the wind's force is kept.

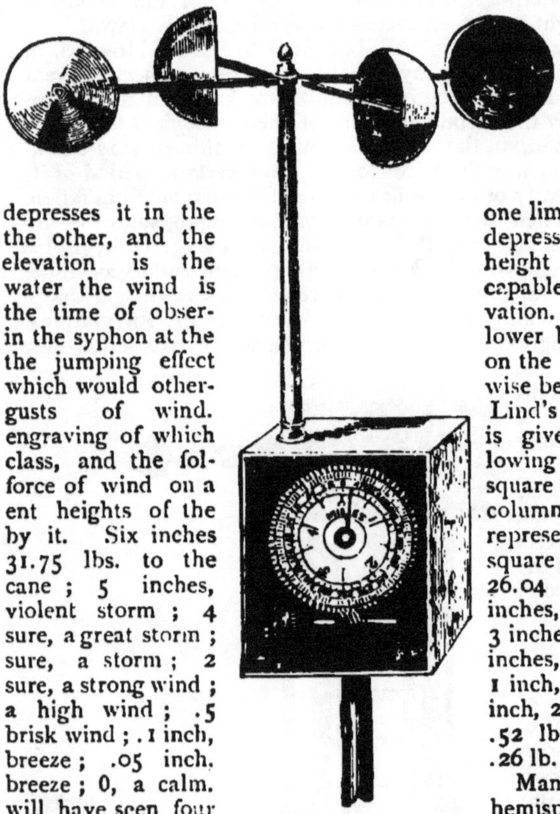

WIND GAUGE.

The cups (measuring from their centres) revolve with one-third of the wind's velocity, and due allowance is made in graduating the indicating dials, so that the true velocity is obtained by direct observation.

JANUARY COLD SNAPS AT TORONTO.

A Toronto correspondent writes: "As it may be interesting to some of the readers of your Almanac, whether as hints for probabilities, or as a mere retrospect of past Januaries, I give the following records of cold snaps, under which head I include all those periods in which the thermometer ranged at, near or below zero, at 8 o'clock a.m.:

January of year.
1855—23, 24, 25.
1856—8, 9, 10, 11; 20, 21; 25, 26.
1857—6, 7, 8; 11, 17, 18, 22, 25.
1858—Up to 27th very little ice in Bay; frozen on 31st.
1859—8, 9, 10, but Bay open on 28th.
1860—1st (10° below), then mild to 31st, 4° below.
1861—10, 11, 12, 13; Bay frozen.
1862—3, 4, 5; 13, 14, 15; but Bay cleared on 15th.
1863—Bay clear till 16th; 17th 8° below zero.
1864—8, 9; 21, 22.
1865—7, 8, 9; 16, 17, 18, 19.
1866—4, 5, 6, 7, 8, 9.

January of year.
1867—15, 16, 17, 18, 19; 29.
1868—12, 13, 14; 22; 27, 28, 29.
1869—22, 25; mild up to 18th; saw a pansy plucked in bloom in open air on 30th.
1870—9; 14; rest of month mild.
1871—7; 22, 24, 25.
1872—7, 8; 29.
1873—18, 25, 29.
1874—Mild or moderate till 30th; 2° below zero.
1875—10, 11, 12; followed by coldest February known here.
1876—Mild or moderate all through.
1877—3, 8, 9, 12; so far gloomy.

"I observe from my records that our January cold snaps have almost invariably been soon followed by snow-storms. That it 'always moderates to snow' is an old and very true weather proverb, which might long ago have dissipated the delusion that the source of these supplies of moisture is to the eastward. Were this the fact, we should see very little moderation of temperature resulting, and, indeed, we should see very little precipitation of either snow or rain, for the North Atlantic gives off a very inadequate quantity to meet our wants, and long before that little could reach us, driven by a north-east or east wind, it would be all appropriated by Labrador, Eastern Canada, or New England. But almost everybody believes our snow and rain storms come from eastward because they come with eastern winds. The stratum of east wind is, however, a mere surface current—an incident of the storm—not the real snow or rain carrier; that runs above the eastern wind, and almost always in an opposite direction. I have, from a most advantageous position, watched the ingress of scores of rain and snow storms, but never yet have I detected one advancing from eastward; yet when they have fully set in, ninety-nine persons out of every hundred would swear they had come from the eastward. But if this were the fact, why should it snow at Detroit or Cleveland before Toronto? Why at Toronto before Kingston or Montreal, and at Montreal before Quebec? I have anticipated many a rain and snow storm by twelve, twenty-four, or even thirty-six hours, warned by the persistent eastern fore-blow; and when at length they approached, I have seen them crossing from the Grimsby shore over the lake and landing on our side; but having once begun to fall on us, the flakes or drops have all obeyed the force of the surface current through which they fell from the upper cloud field,

"For a very sufficient reason 'it always moderates to snow.' It would, however, be more correct to say the snow brings the moderation. It is formed from vapour sent to us from warm regions, and where evaporation is ever active and abundant. When this vapour is carried over a frozen region, or arrives at one, it falls in the form of snow; when a contrary condition exists it falls in rain.

"Many of our winter cold snaps are of only brief duration. In the course of twenty-four, or even twelve hours, we may see the thermometer plunge down 35 or 40 degrees, as it did on the night of January 11th, 1877. Sometimes, when it thus precipitately descends, it as rapidly re-ascends, even to a higher point than it held before.

TWO MILD FEBRUARIES—1877 AND 1834.

It has occurred to me that many of your readers would be interested in an account of the winter of 1833-4, more particularly as the month of February of that season was even milder, in some respects, than the one which has just passed. The fall of 1833 was pleasant, and ploughing was not entirely stopped by frost until about the 20th of November. Steady frost set in about that time, and snow fell in sufficient quantity on the 6th December to make sleighing. The fore part of the winter was steady but mild, and the St. Lawrence was not frozen over so that teams could cross until the 18th of January. There was one week of cold weather, the thermometer reaching to 18° below zero. February set in mild, and there was almost constant thaw until near the end of the month, and in the third week snow had nearly disappeared. There were two thunder-storms in February. One on the 20th prevailed over the Province, and did much damage. On the 24th crows and flies made their appearance. On the 26th there was a fall of snow sufficient to make sleighing until the 4th of March. Mild weather again set in, and the frost began to leave the ground. On the 13th we began to prepare ground for a garden by removing stones and stumps. Green blades of grass began to appear, and sheep and young cattle found themselves food, and the rivers also were free of ice. On the 20th March there was a severe thunder-storm with heavy rain, which raised the rivers. On the 21st frost again set in, and on the 25th there was a fall of snow, which lay a day or two. Warm weather again set in on the 30th, and on the 3rd of April frogs were heard croaking for the first time, and the woods began to be enlivened by the music of the birds. On the 1st April the steamboat "Franklin" commenced running from St. Johns to Lake Champlain. The steamer "Chateauguay" also began her regular trips from Chateauguay Basin to Lachine early in April. Ploughing was now general, and some were sowing. Mosquitos made their appearance about the 7th, and the weather was so warm that fires were not needed. The roads became dry and good, and all the month of April continued fine. There was thunder several times, and also a few flurries of snow. By the 21st vegetation was far advanced, and many trees were nearly in full leaf—even the maple; wheat was above ground, and pools of water full of tadpoles. In the first week of May there was thunder and occasionally frost. On the 13th there was snow, and on the 14th it snowed for ten hours. On the morning of the 15th there was severe frost, ice on pools being about half an inch thick. No harm resulted from the frost, as the snow protected vegetation. It was a dry, hot summer, and an early harvest, but the crop was tolerably good.

F. W. S.

Huntington, 15th March, 1877.

GENERAL RULES AND LAWS FOR STORMS, WIND AND WEATHER.—1st. For all meteoric observations on large bodies of land in the temperate zones fix yourself with face looking towards the main sea or ocean. 2nd. If the wind is blowing on your face at the rate of only four to eight miles an hour, rain is not apt to be present, as the atmosphere will likely be expanding and absorbing water and converting it into vapor, and holding it suspended in the air until the proper planetary phenomena transpire. 3rd. If the wind is moving at the rate of ten to fifteen miles an hour it is probably passing to a rain, hail or snow-storm farther inland, and you will likely find the barometer falling. 4th. If it is moving at the rate of twenty to thirty miles an hour, it is likely passing further inland to a more violent rain, hail or snow precipitation. If it is moving at forty to sixty miles an hour it is probably blowing further inland to a waterspout, hurricane or tornado, etc. 5th. When you find the wind turned round (about 180 degrees) and blowing on your back, the storm has passed you, and it is on its way towards the sea, and you will be apt to find the barometer rising. 6th. If the wind turns partially around (about 90 degrees) and blows on your right and off your left, the storm is then supposed to pass by your left on its way towards the sea. If the wind should turn the other way around (about 90 degrees) and blow on your left and off your right, then the storm is likely passing by the way of your right towards the sea. The wind often moves from all quarters to the point of precipistation. There is frequently a kind of short lull in front of travelling storms. There are mild local rains caused by slow changing positions of the planets, etc. By noting these rules and watching the barometer the course and severity of storms may soon be tolerably well understood, even for great distances from the observer.

FORESTS AND RAINFALL.—The relation between woodlands and rain-fall and other climatic conditions has of late been the subject of much dogmatic theorizing. A comparison of maps in Walker's "Statistical Atlas of the United States" shows that the forests in Washington Territory have an annual rain-fall of 60 inches and upward. The magnificent forests found from Minnesota to Maine have a rain-fall precisely identical with that of the nearly treeless prairies which extend westward from Chicago, viz., from 28 to 40 inches. The northern part of the Michigan Peninsula, with its heavy timber, is marked with precisely the same rain-fall as large portions of southern Minnesota, lying in the same latitudes and nearly treeless.

THE MOON AND THE WEATHER.

The notion that the moon exerts an influence on weather is so deeply rooted that, notwithstanding all the attacks which have been made against it since meteorology has been seriously studied, it continues to retain its hold upon many of us. And yet there never was a popular superstition more utterly without a basis than this one. If the moon did really possess any power over weather, that power could only be exercised in one of three ways—by reflection of the sun's rays, by attraction, or by emanation. No other form of action is conceivable. Now, as the brightest light of a full moon is never equal in intensity or quantity to that which is reflected towards us by a white cloud on a summer day, it can scarcely be pretended that weather is affected by such a cause. That the moon does exert attraction on us is manifest. We see its working in the tides; but though it can move water it is most unlikely that it can do the same to air, for the specific gravity of the atmosphere is so small that there is nothing to be attracted. Laplace calculated, indeed, that the joint attraction of the sun and moon together could not stir the atmosphere at a quicker rate than five miles a day. As for lunar emanations, not a sign of them has ever been discovered. The idea of an influence produced by the phases of the moon is therefore based on no recognizable cause whatever. Furthermore, it is now distinctly shown that no variations at all really occur in weather at the moment of the changes of quarter any more than at other ordinary times. Since the establishment of meteorological stations all over the earth, it has been proved by millions of observations that there is no simultaneousness whatever between the supposed cause and the supposed effect. The whole story is a fancy and a superstition, which has been handed down to us uncontrolled, and which we have accepted as true because our forefathers believed it. The moon exercises no more influence on weather than herrings do on the government of Switzerland.—*Blackwood's Magazine.*

WEATHER WISDOM.—"Many persons are predicting an early, long and severe winter. The indications are the unusual abundance of pine cones, the big piles of dirt the gophers are making about their holes, the unusual thickness of the corn shucks, the industry of the woodpeckers in laying up stores of acorns, the early rising of the springs in the mountains, and the mildness of the weather last winter." So reads a paragraph in *Forest and Stream* for October, at which we are inclined to laugh heartily. Such indications may read any way. Who has ever seen woodpeckers gathering acorns? These birds live entirely upon insects, and keep on the wing searching for them all winter. The thickness or thinness of the shucks on the corn depends entirely upon the amount of sun and moisture the grain has had during the summer, and has nothing to do with the weather that is to come long after the harvest has been gathered indoors. As to last winter (1877) being a mild one, we would answer so was the one preceding it. The poor little gopher we know nothing of in Canada, but our woodchucks are as yet making but little preparation for the winter.

THE OREGON EARTHQUAKE OF 1877.

The Portland *Oregonian* of Oct. 13th gives an account of an earthquake which took place the previous afternoon. At 1.53 o'clock a distinct earth shock, followed in a few seconds by another and severer one, passed over the city from north to south. It was not, as is usually the case, preceded by premonitory grumblings, but came with terrifying suddenness. The scene on the principal streets, as the people became conscious of the cause of the agitation, was one of the wildest confusion and, for a moment, of terror. From houses and stores people with white, scared faces rushed into the streets, cigars dropped from the mouths of smokers, horses snorted and dogs whined with fear, the air, as well as the earth, seemed filled with a mysterious and awful power—the streets seemed turned into a mad carnival of fear. This was for one moment; the next everybody was trying to convince everybody else that he "wasn't a bit scared." While the shock was very severe, or at least seemed so to Oregonians, it was not accompanied by loss of life or destruction of property to any great extent. A panic was created at each of the public schools, and children made for the open air without considering the manner of their going. At the North building they rushed pell-mell down stairs, and in the turmoil several children were badly bruised. At the Central and High schools a similar occurrence took place, and in Harrison street school the terror of the children was awful. Several windows were broken, and it seemed as though the house would certainly fall. The shock was much harder in the southern part of the city, and many residences were well shaken up. In the county jail, several feet below the surface, it was very severe, and a stove was knocked from its "moorings" and thrown over. This earthquake extended over a considerable territory.

WEATHER FORECASTS.—Speaking of weather predictions, Dr. Johnson is reported to have said that a weather-wise man might in the morning foretell what sort of weather there would be between that time and evening, but that he was powerless to predict the weather four-and-twenty hours in advance. Till recently this observation remained substantially true; but now, owing to the invention of the electric telegraph, and its ramification over the civilized world, the statement needs to be received with some modification. Formerly the meteorologist's field of observation was bounded by the limited horizon visible to his own eyes; whereas now he may sit in his office in London and receive from a staff of messengers, travelling more swiftly than the swiftest hurricane, a series of simultaneous reports from places as far apart as Bergen, Gibraltar, the Texel, and Valentia. The practical result of these improvements is that it is now possible to issue weather warnings to seafaring persons and others, which are in the majority of cases absolutely verified; and to this cause it is no doubt in some degree attributable that the remarkable storm which swept over these islands on the night of Sunday last, and which did so much damage on land, caused, in comparison with its exceeding violence, little damage by sea.

Mariners, having received timely warning, either sought shelter at once, or forebore from quitting the shelter in which they already lay. In enumerating some of the places from which a modern meteorologist receives his almost instantaneous reports, we purposely omitted to mention America, because it is doubtful whether, with our present stock of knowledge, any practical inferences of value can be drawn from the weather phenomena of a region separated from us by an ocean nearly three thousand miles wide. The storm from New York, which was predicted to reach our shores on the 10th inst., certainly did not visit us on that day, which was remarkably calm; but at the same time it is possible that the great gale of Sunday night was part of the same convulsion, although it had been delayed four days on the road.—*London Graphic, Oct.*

THE ST. PATRICK'S DAY COLD DIP.

An Elora paper says: "Vennor was safe in predicting a cold dip in the neighborhood of St. Patrick's Day. A gentleman in Toronto who makes weather notes informs us that a fall in the thermometer to the zero line has occurred about the date of the equinox in nearly every one of the last twenty-three years, and a gentleman in Elora gives us the following markings of lowest thermometer in March since 1869:

1869—March	16,	5° below zero.	1874—March	12,	3° below zero.
1870— "	19,	4° "	*1875— "	23,	10° "
1871— "	19,	8° "	1876— "	18,	5° "
1872— "	20,	20° "	1877— "	17,	16° "

*Below zero 10th, 19th, 21st and 22nd.

Modern Philosophers and Lightning Rods.—Leading philosophers comment on and teach the theory of the earth (sun and other planets) as parting with heat by radiation. The lightning rod vendor fixes the point of his conductor in the air over the building that is to be protected, and locates the other end of the rod in the earth beneath to receive the lightning or electric discharge from the air above. By this mode we see the lightning rod man acting consistently with the laws of nature, but he does not appear to understand for what reason he does this; while, to harmonize with the theory of modern philosophy, he should invert his conductors by setting the points that are to collect the wild electricity in the earth, and spread fan-like radiators above or over the buildings at the other end of the rod to radiate the electricity from the earth and houses into space. If the earth parted with its heat loosely by radiation, the heat liberated from the clouds by the condensation of the vapours should pass out into space. But this is not nature's plan. The electricity is absorbed by the solid earth from the vapor, though it does not appear to be understood this way as yet. But time does much.—*Prof. Mansill.*

USE OF BIRDS TO THE FARMER.—The swallow, swift and nighthawk are the guardians of the atmosphere. They check the increase of insects that otherwise would overload it. Woodpeckers, creepers and chicadees are the guardians of the trunks of trees; warblers and flycatchers protect the foliage; blackbirds, thrushes, crows and larks protect the surface of the soil; snipe and woodcock the soil under the surface. Each tribe has its respective duties to perform in the economy of nature, and it is an undoubted fact that if the birds were all swept off from the earth man could not live upon it; vegetation would wither and die, and insects would become so numerous that no living thing could withstand their attacks. The wholesale destruction occasioned by the grasshoppers which have lately devastated the West is undoubtedly caused by the thinning out of the birds, such as grouse, prairie-hens, &c., which feed upon them. The great and inestimable service done to the farmer, gardener and florist by the birds is only becoming known by sad experience. Spare the birds and save your fruit; the little corn and fruit taken by them is more than compensated by the vast quantities of noxious insects destroyed. The long persecuted crow has been found, by actual experiment, to do far more good by the immense numbers of insects he devours than the little harm he does by by the few grains of corn he pulls up. He is one of the farmer's best friends.

AËROLITES—METEORIC STONES.—Unlike falling stars, which become extinguished in the upper regions without noise, and without leaving any trace of their existence, aërolites, or meteoric stones, reach the surface of the earth. These have been met with in all parts of the world, and in no one place more than another. They move with a great velocity, and shine with an intensely bright light. They generally strike the earth in an oblique direction, and frequently with such force as to bury themselves many feet in the soil. When they first fall they are so hot that they have been known to turn the sand in which they bury themselves into glass, coating the hole so as to form a tube which could be taken out entire. They have been seen to explode at the height of thirty and forty miles. Meteoric stones are of various shapes and sizes. They have always rough edges and depressions in their surfaces, and are coated with a black, shining crust, but of a grayish color within. One that fell in the township of Madoc, Hastings county, Ont., weighed pounds; this specimen is now in the Geological Museum at Montreal. Another, which fell in South America, weighed 30,000 pounds. One in Arkansas weighs 1,635 pounds; still another weighs 14,000 pounds. The most remarkable masses of meteoric iron occur in the district of Chaco-Gualamba, in South America, where there is one whose weight is estimated at 30,000 pounds. Besides nickel, which sometimes amounts to nearly 20 per cent., meteoric iron often contains small percentages of cobalt, tin, copper, and manganese, and not unfrequently nodules of magnetic iron pyrites are embedded in the mass. Meteoric iron is perfectly malleable, and may be worked like manufactured iron. Specimens of native iron, intimately mixed with rock, have been found at Portage du Fort, on the Ottawa river, but whether these represent terrestrial native iron or iron ore altered by artificial means is not yet satisfactorily determined. Many theories have been propounded to account for the origin of meteoric stones. Laplace supposed that they were projected from volcanoes in the moon, and falling within the attraction of the earth, were drawn to its surface ; but this does not account for their great velocity. Moreover, from the large number falling on the earth, the moon itself would soon be reduced to a mere meteor, and fall like the rest. Others have given them a terrestrial origin. Chladni supposes that they are small bodies circulating around the sun, which coming within the earth's attraction, are drawn to it, and become heated and ignited by the friction occasioned by their rapid motion through our atmosphere. The previously dark and invisible meteor becomes luminous, and the ignited and incandescent particles becoming detached from the main mass form a glowing train of light behind the aërolite. This is the theory now generally received, and which more nearly accounts for all their phenomena.

RAIN-FALL AND SOLAR SPOTS.—Of late there has been some interest shown in the supposed relation between the periodicity of rainfall and the periodicity of solar spots. The researches of scientists seem to show that there is a very close connection between solar disturbances and terrestrial phenomena ; a marked correspondence being observed between magnetic and electric disturbances on the earth and the occurrence of spots on the sun. A periodicity of cyclones in the Indian Ocean is also connected with a similar periodicity of solar spots. A corresponding change of atmospheric temperature and solar spots has also been noted, and it has been found that more rain falls in years of maxima solar spots than in minima solar spot years, showing that here, as well as everywhere else, the sun exerts an influence.

GOOSE-BONE WEATHER PREDICTIONS.—The goose-bone is more closely watched in Kentucky than in any other part of the country. It has been handed down among the early traditions of the State, and may be called the Kentucky weather prophet. It is to be found in most Kentucky country homes, and in many parts of the State the farmers consult it, and prepare for handling their crops in accordance with its readings. The prophecy of the goose-bone does not extend beyond the year in which the goose was hatched, and the prediction is for the three winter months only. Take the breast bone of a last spring's goose and divide it into three equal parts, and the different divisions will represent December, January and February. The breast bone of a goose is translucent, and if clear when held up to the light, the weather will be mild and pleasant ; but if covered with cloud-like blots it will be gloomy and cold ; the heavier the blots the colder will be the weather.

A study of this year's goose-bono indicates that the weather for

December will be cloudy and gloomy, probably with much rain and snow, not very cold, but, withal, a very disagreeable month. About the last of December we shall have some cold weather, which will continue to grow colder as January advances. The month of January will be a cold one throughout, with some very severe weather during the last part of the month. On the prophecy of the goose-bone, it may be predicted that about the last of January we will have the coldest weather experienced for a number of years. February will be more pleasant and spring-like, betokening an early return of the flowers. During the last of the month, however, there will be a few cold days, but no severe weather.

Such is the prophecy of the goose-bone, and as we have the word of a good old farmer up in Woodford county that it has not failed for fifty years, we may as well prepare to meet it, and need not be surprised if we have good skating on the Ohio river during the latter part of January.—*Louisville Commercial*, 1876.

HAIL.—As a general rule, hailstorms occur at the close of long periods of calm, hot and sultry weather. They are immediately preceded by a fall of the barometer, and, what is unusual before rain, a corresponding fall of the thermometer. The thermometer, during a hailstorm, has been known to sink through 77° Fahr. As a rule, hail clouds are not so high as rain clouds, while the area of hailstorms frequently extends over great distances in a linear direction. Their breadth is seldom very great. The motion is rapid, often forty miles an hour. A peculiar rustling sound in the air often precedes the fall of hail. This is accompanied by a darkness similar to a total eclipse of the sun. Hailstorms are seldom of long duration; from three or four minutes to a quarter of an hour is the usual limit.

DROUGHTS AND RAINS.—During long, hot, dry terms of weather in medium latitudes, or at distances from the sea, the vapors are carried over or past them, and condensed in the more northern latitudes, or farther from the ocean. Medium latitudes from the sea receive a fair share of rain in moderate seasons. During long cold intervals the vapors are in all likelihood condensed before reaching far north or inland from the ocean.—*Mansill's Almanac.*

THINGS NOT GENERALLY KNOWN.—Caterpillars never produce young; flies, bees, etc., never grow larger after their escape from the cocoon. Most people suppose that the little flies that we see around are the same kind as the large ones, only younger; but the fact is that they are the same size as when hatched out from the cocoon.

BATHS !

—:o:—

Electric, Sulphur and Turkish Baths

AT

A. NORMAN'S, No. 4 Queen St. East, Toronto.

GALVANIC AND MAGNETIC BELTS, &c.

Great Devonshire Cattle Food

May be relied on as containing no *copperas* or other metallic substance, and is unquestionably the only scientific combination to produce a healthy, salable and working condition in horses, and fattening cattle to a degree produced by no other feeder, shown by the successful *exportation* of cattle to England fed with the GREAT DEVONSHIRE FOOD.

☞Ask for the Devonshire and take no other.

Beware of Worthless Imitations. *Price* $1.00 *per Box.*

SOLE MANUFACTURER,

JOHN LUMBERS, 101 Adelaide St. East, Toronto.

Being extensively engaged in

Growing Seeds

we can offer special inducements to the trade.

FARMERS and others in the back settlements can get their Seeds by Mail.

—

Catalogues Free.

OUR SKATE PREMIUMS.

THE EUREKA SKATES.

A pair of these skates, worth $4.00, will be sent to every person who obtains $15 in new subcriptions to the WITNESS publications, deducting no commission therefrom.

A pair of EUREKA SKATES, worth $2.75, will be sent to every person who obtains $10 in new subscriptions to the WITNESS publications, deducting no commission therefrom.

THE CANADIAN CLUB SKATE.

A pair of these skates, worth $2.75, will be given to every person who obtains $9 in new subscriptions to the WITNESS publications, deducting no commission therefrom.

SAMPLE copies and necessary instructions sent free on application to the publishers

JOHN DOUGALL & SON, MONTREAL.

GIRLS AND BOYS.

This premium has met with great favour all over Canada. More than a thousand girls and boys are working for the skates. The following is what those who got them last year say:

"I am delighted with them."

"The boys that have seen them like them very much."

"I received the skates all right, and am much pleased with them; they exceed my expectations."

"The skates are complete in every respect."

WHILE EVERYBODY CAN GET SUCH A PAIR OF SKATES AS THESE SO EASILY, WHO WILL GO WITHOUT ANY?

Send to John Dougall & Son, Montreal, for sample copies, lists, &c.

www.ingramcontent.com/pod-product-compliance
Lightning Source LLC
Chambersburg PA
CBHW021817190326
41518CB00007B/635

* 9 7 8 3 3 3 7 2 5 7 8 2 8 *